LONDON MATHEMATICAL SOCIETY LE

Managing Editor: Professor J.W.S. Cassels, Departm(
University of Cambridge, 16 Mill Lane, Cambridge C

The books in the series listed below are available from ...., or, in case of difficulty, from Cambridge University Press.

- 34 Representation theory of Lie groups, M.F. ATIYAH et al
- 36 Homological group theory, C.T.C. WALL (ed)
- 39 Affine sets and affine groups, D.G. NORTHCOTT
- 46 p-adic analysis: a short course on recent work, N. KOBLITZ
- 49 Finite geometries and designs, P. CAMERON, J.W.P. HIRSCHFELD & D.R. HUGHES (eds)
- 50 Commutator calculus and groups of homotopy classes, H.J. BAUES
- 57 Techniques of geometric topology, R.A. FENN
- 59 Applicable differential geometry, M. CRAMPIN & F.A.E. PIRANI
- 66 Several complex variables and complex manifolds II, M.J. FIELD
- 69 Representation theory, I.M. GELFAND et al
- 74 Symmetric designs: an algebraic approach, E.S. LANDER
- 76 Spectral theory of linear differential operators and comparison algebras, H.O. CORDES
- 77 Isolated singular points on complete intersections, E.J.N. LOOIJENGA
- 79 Probability, statistics and analysis, J.F.C. KINGMAN & G.E.H. REUTER (eds)
- 80 Introduction to the representation theory of compact and locally compact groups, A. ROBERT
- 81 Skew fields, P.K. DRAXL
- 82 Surveys in combinatorics, E.K. LLOYD (ed)
- 83 Homogeneous structures on Riemannian manifolds, F. TRICERRI & L. VANHECKE
- 86 Topological topics, I.M. JAMES (ed)
- 87 Surveys in set theory, A.R.D. MATHIAS (ed)
- 88 FPF ring theory, C. FAITH & S. PAGE
- 89 An F-space sampler, N.J. KALTON, N.T. PECK & J.W. ROBERTS
- 90 Polytopes and symmetry, S.A. ROBERTSON
- 91 Classgroups of group rings, M.J. TAYLOR
- 92 Representation of rings over skew fields, A.H. SCHOFIELD
- 93 Aspects of topology, I.M. JAMES & E.H. KRONHEIMER (eds)
- 94 Representations of general linear groups, G.D. JAMES
- 95 Low-dimensional topology 1982, R.A. FENN (ed)
- 96 Diophantine equations over function fields, R.C. MASON
- 97 Varieties of constructive mathematics, D.S. BRIDGES & F. RICHMAN
- 98 Localization in Noetherian rings, A.V. JATEGAONKAR
- 99 Methods of differential geometry in algebraic topology, M. KAROUBI & C. LERUSTE
- 100 Stopping time techniques for analysts and probabilists, L. EGGHE
- 101 Groups and geometry, ROGER C. LYNDON
- 103 Surveys in combinatorics 1985, I. ANDERSON (ed)
- 104 Elliptic structures on 3-manifolds, C.B. THOMAS
- 105 A local spectral theory for closed operators, I. ERDELYI & WANG SHENGWANG
- 106 Syzygies, E.G. EVANS & P. GRIFFITH
- 107 Compactification of Siegel moduli schemes, C-L. CHAI
- 108 Some topics in graph theory, H.P. YAP
- 109 Diophantine analysis, J. LOXTON & A. VAN DER POORTEN (eds)
- 110 An introduction to surreal numbers, H. GONSHOR
- 111 Analytical and geometric aspects of hyperbolic space, D.B.A. EPSTEIN (ed)
- 113 Lectures on the asymptotic theory of ideals, D. REES
- 114 Lectures on Bochner-Riesz means, K.M. DAVIS & Y-C. CHANG
- 115 An introduction to independence for analysts, H.G. DALES & W.H. WOODIN
- 116 Representations of algebras, P.J. WEBB (ed)
- 117 Homotopy theory, E. REES & J.D.S. JONES (eds)
- 118 Skew linear groups, M. SHIRVANI & B. WEHRFRITZ
- 119 Triangulated categories in the representation theory of finite-dimensional algebras, D. HAPPEL
- 121 Proceedings of *Groups - St Andrews 1985*, E. ROBERTSON & C. CAMPBELL (eds)
- 122 Non-classical continuum mechanics, R.J. KNOPS & A.A. LACEY (eds)
- 124 Lie groupoids and Lie algebroids in differential geometry, K. MACKENZIE
- 125 Commutator theory for congruence modular varieties, R. FREESE & R. MCKENZIE
- 126 Van der Corput's method of exponential sums, S.W. GRAHAM & G. KOLESNIK
- 127 New directions in dynamical systems, T.J. BEDFORD & J.W. SWIFT (eds)
- 128 Descriptive set theory and the structure of sets of uniqueness, A.S. KECHRIS & A. LOUVEAU
- 129 The subgroup structure of the finite classical groups, P.B. KLEIDMAN & M.W.LIEBECK
- 130 Model theory and modules, M. PREST

131	Algebraic, extremal & metric combinatorics, M-M. DEZA, P. FRANKL & I.G. ROSENBERG (eds)
132	Whitehead groups of finite groups, ROBERT OLIVER
133	Linear algebraic monoids, MOHAN S. PUTCHA
134	Number theory and dynamical systems, M. DODSON & J. VICKERS (eds)
135	Operator algebras and applications, 1, D. EVANS & M. TAKESAKI (eds)
136	Operator algebras and applications, 2, D. EVANS & M. TAKESAKI (eds)
137	Analysis at Urbana, I, E. BERKSON, T. PECK, & J. UHL (eds)
138	Analysis at Urbana, II, E. BERKSON, T. PECK, & J. UHL (eds)
139	Advances in homotopy theory, S. SALAMON, B. STEER & W. SUTHERLAND (eds)
140	Geometric aspects of Banach spaces, E.M. PEINADOR and A. RODES (eds)
141	Surveys in combinatorics 1989, J. SIEMONS (ed)
142	The geometry of jet bundles, D.J. SAUNDERS
143	The ergodic theory of discrete groups, PETER J. NICHOLLS
144	Introduction to uniform spaces, I.M. JAMES
145	Homological questions in local algebra, JAN R. STROOKER
146	Cohen-Macaulay modules over Cohen-Macaulay rings, Y. YOSHINO
147	Continuous and discrete modules, S.H. MOHAMED & B.J. MÜLLER
148	Helices and vector bundles, A.N. RUDAKOV et al
149	Solitons, nonlinear evolution equations & inverse scattering, M. ABLOWITZ & P. CLARKSON
150	Geometry of low-dimensional manifolds 1, S. DONALDSON & C.B. THOMAS (eds)
151	Geometry of low-dimensional manifolds 2, S. DONALDSON & C.B. THOMAS (eds)
152	Oligomorphic permutation groups, P. CAMERON
153	L-functions and arithmetic, J. COATES & M.J. TAYLOR (eds)
154	Number theory and cryptography, J. LOXTON (ed)
155	Classification theories of polarized varieties, TAKAO FUJITA
156	Twistors in mathematics and physics, T.N. BAILEY & R.J. BASTON (eds)
157	Analytic pro-p groups, J.D. DIXON, M.P.F. DU SAUTOY, A. MANN & D. SEGAL
158	Geometry of Banach spaces, P.F.X. MÜLLER & W. SCHACHERMAYER (eds)
159	Groups St Andrews 1989 volume 1, C.M. CAMPBELL & E.F. ROBERTSON (eds)
160	Groups St Andrews 1989 volume 2, C.M. CAMPBELL & E.F. ROBERTSON (eds)
161	Lectures on block theory, BURKHARD KÜLSHAMMER
162	Harmonic analysis and representation theory for groups acting on homogeneous trees, A. FIGA-TALAMANCA & C. NEBBIA
163	Topics in varieties of group representations, S.M. VOVSI
164	Quasi-symmetric designs, M.S. SHRIKANDE & S.S. SANE
165	Groups, combinatorics & geometry, M.W. LIEBECK & J. SAXL (eds)
166	Surveys in combinatorics, 1991, A.D. KEEDWELL (ed)
167	Stochastic analysis, M.T. BARLOW & N.H. BINGHAM (eds)
168	Representations of algebras, H. TACHIKAWA & S. BRENNER (eds)
169	Boolean function complexity, M.S. PATERSON (ed)
170	Manifolds with singularities and the Adams-Novikov spectral sequence, B. BOTVINNIK
171	Squares, A.R. RAJWADE
172	Algebraic varieties, GEORGE R. KEMPF
173	Discrete groups and geometry, W.J. HARVEY & C. MACLACHLAN (eds)
174	Lectures on mechanics, J.E. MARSDEN
175	Adams memorial symposium on algebraic topology 1, N. RAY & G. WALKER (eds)
176	Adams memorial symposium on algebraic topology 2, N. RAY & G. WALKER (eds)
177	Applications of categories in computer science, M. FOURMAN, P. JOHNSTONE, & A.. PITTS (eds)
178	Lower K- and L-theory, A. RANICKI
179	Complex projective geometry, G. ELLINGSRUD, C. PESKINE, G. SACCHIERO & S.A. STRØMME (eds)
180	Lectures on ergodic theory and Pesin theory on compact manifolds, M. POLLICOTT
181	Geometric group theory I, G.A. NIBLO & M.A. ROLLER (eds)
182	Geometric group theory II, G.A. NIBLO & M.A. ROLLER (eds)
183	Shintani zeta functions, A. YUKIE
184	Arithmetical functions, W. SCHWARZ & J. SPILKER
185	Representations of solvable groups, O. MANZ & T.R. WOLF
186	Complexity: knots, colourings and counting, D.J.A. WELSH
187	Surveys in combinatorics, 1993, K. WALKER (ed)
189	Locally presentable and accessible categories, J. ADAMEK & J. ROSICKY
190	Polynomial invariants of finite groups, D.J. BENSON
191	Finite geometry and combinatorics, F. DE CLERCK et al
192	Symplectic geometry, D. SALAMON (ed)
196	Microlocal analysis for differential operators, A. GRIGIS & J. SJÖSTRAND
197	Two-dimensional homotopy and combinatorial group theory, C. HOG-ANGELONI, W. METZLER & A.J. SIERADSKI (eds)
198	The algebraic characterization of geometric 4-manifolds, J.A. HILLMAN

London Mathematical Society Lecture Note Series. 196

Microlocal Analysis for Differential Operators
An Introduction

Alain Grigis
Université de Paris Nord

Johannes Sjöstrand
Université de Paris Sud

Published by the Press Syndicate of the University of Cambridge
The Pitt Building, Trumpington Street, Cambridge CB2 1RP
40 West 20th Street, New York, NY 10011-4211, USA
10 Stamford Road, Oakleigh, Melbourne 3166, Australia

© Cambridge University Press 1994

First published 1994

Printed in Great Britain at the University Press, Cambridge

Library of Congress cataloguing in publication data available

British Library cataloguing in publication data available

ISBN 0 521 44986 3 paperback

Transferred to digital reprinting 2000
Printed in the United States of America

Contents

Introduction	1
1. Symbols and oscillatory integrals	5
2. The method of stationary phase	19
3. Pseudodifferential operators	27
4. Application to elliptic operators and L^2 continuity	41
5. Local symplectic geometry I (Hamilton–Jacobi theory)	55
6. The strictly hyperbolic Cauchy problem — construction of a parametrix	67
7. The wavefront set (singular spectrum) of a distribution	77
8. Propagation of singularities for operators of real principle type	89
9. Local symplectic geometry II	97
10. Canonical transformations of pseudodifferential operators	107
11. Global theory of Fourier integral operators	119
12. Spectral theory for elliptic operators	131
Bibliography	145

0 Introduction

This book corresponds to a graduate course which has been taught many times at the universities Paris-Sud, Paris-Nord and Paris 7, and other places. The aim of this text is to give the foundation of what is nowadays called microlocal analysis in the C^∞ framework, as it was created in the sixties and seventies by Kohn–Nirenberg, Maslov and Hörmander. Our presentation follows essentially the one given by Hörmander [Hö2]; as for the symplectic geometry, we have been inspired by the lecture notes of Duistermaat [D].

This subject is of growing importance, with a range of applications going beyond the original problems of linear partial differential equations. In particular the link with quantum mechanics is now firmly established, and there is a growing number of books covering more or less specialized parts of the theory. We believe that a short monograph concentrating on the basic principles could be of value, not only for the graduate student, but also for the mathematician who wants to get *quickly* into the subject and to understand its basic mechanisms. For this, the classical PDE framework seemed to us the most suitable one. The basic principles of microlocal analysis are essentially only two: integration by parts and the method of stationary phase. Compared with the article [Hö2] and many of the other presentations, we have insisted even more on the stationary phase method, which appears already in the development of the theory of pseudodifferential operators and also (as in Melin–Sjöstrand [MS]) in the proof of the equivalence of phase functions in the global theory of Fourier integral operators.

Readers of this book are expected to be familiar with the theory of distributions, in particular with the Fourier transform. They should also know some basic functional analysis and differential geometry. The following is a brief outline of the contents of the book :

1) Symbols and oscillatory integrals. Here we introduce some notions related to asymptotic developments and the "non-stationary phase lemma". We also study some properties of Fourier integral operators and distributions.

2) The method of stationary phase. In this chapter we prove the Morse lemma on the normal form of C^∞ functions near a non-degenerate critical point, and apply it in order to find the asymptotic expansion of an integral of the form $\int e^{i\lambda\phi(x)}a(x)dx$, when $\lambda \longrightarrow \infty$ and ϕ has critical points.

3) Pseudodifferential operators. Here we develop the theory of these operators and its symbolic calculus with composition, adjoints and changes of variables.

4) Applications to elliptic operators and L^2 continuity. We construct approximate inverses (parametrices) to elliptic operators, which leads us naturally to leave the class of differential operators. We also study the action of pseudodifferential operators in L^2 and Sobolev spaces.

5) Local symplectic geometry I (Hamilton–Jacobi theory). We recall briefly some basic notions of differential geometry and explain how to solve Hamilton–Jacobi equations locally.

6) The strictly hyperbolic Cauchy problem – construction of a parametrix. We follow the method of P.D. Lax [L], which is a particular case of the general so-called WKB method.

7) The wavefront set (singular spectrum) of a distribution. This is a refinement of the notion of singular support. It not only describes the set where a distribution is singular but also localizes the frequencies that constitute these singularities.

8) Propagation of singularities for operators of real principal type. In the special case of the wave equation this result says that the singular support of a solution is a union of optical rays. The precise formulation of this result requires the notion of wavefront set.

9) Local symplectic geometry II. We discuss canonical transformations and normal forms.

10) Canonical tranformations of pseudodifferential operators. We prove the Egorov theorem, which states that the conjugation of a pseudodifferential operator by a Fourier integral operator gives a new pseudodifferential operator whose principal symbol is obtained by composition by a canonical transformation. This result often permits one to simplify a given pseudodifferential operator.

11) Global theory of Fourier integral operators. We sketch the global theory after establishing its essential ingredient: two non-degenerate phase functions that generate the same Lagrangian manifold give rise to the same classes of oscillatory integrals.

12) Spectral theory for elliptic operators. We prove a theorem of Hörmander on the Weyl asymptotics with a small remainder for self-adjoint elliptic operators of arbitrary order on a compact manifold.

Exercises can be found at the end of each chapter. Some are just for fun and some indicate important results that we did not put in the main text, such as: pseudodifferential operators on manifolds, the sharp Gårding inequality, the Calderón–Vaillancourt theorem (for symbols of type 0,0), 1-dimensional WKB and the Bohr–Sommerfeld quantization condition and some indication on the Weyl quantization of symbols.

We mention other useful monographs: L. Hörmander [Hö4], F. Treves [Tr], M. Taylor [T], M. Shubin [Sh], J. Chazarain and A. Piriou [ChP], V.P. Maslov and M. Fedoriuk [MaFe], V.P. Maslov [Ma1], S. Alinhac and P. Gérard [AGé], D. Robert [R], J-M. Delort [De].

The subject of this book is marked by the influence of L. Hörmander and one of us has had the privilege of being his pupil during some of the most exciting stages of the development of the theory. Many students have followed our lectures through the years and contributed with stimulating questions and remarks. In particular T. Ramond, F. Klopp and L. Nédelec have been helpful with the exercises. From our wifes and children, we have received support and understanding for this job. Many thanks to all these people.

1 Symbols and oscillatory integrals

We shall use the following notation : $\mathbb{R}$ is the set of real numbers, $\mathbb{N} = \{0, 1, 2, \ldots\}$, $\mathbb{R}^n = \mathbb{R} \times \ldots \times \mathbb{R}$ (n factors), $\dot{\mathbb{N}} = \mathbb{N}\setminus\{0\}$, $\dot{\mathbb{R}}^n = \mathbb{R}^n\setminus\{0\}$.

An element $\alpha = (\alpha_1, \ldots, \alpha_n)$ of $\mathbb{N}^n$ will be called a multi-index and the length of α is the corresponding ℓ^1-norm : $|\alpha| = \alpha_1 + \ldots + \alpha_n$. (For points $x \in \mathbb{R}^n$, we denote by $|x|$ or by $\|x\|$ the ordinary Euclidean norm.) We write $x^\alpha = x_1^{\alpha_1} \ldots x_n^{\alpha_n}$, $x = (x_1, \ldots, x_n)$, $\partial_x^\alpha = \partial_{x_1}^{\alpha_1} \ldots \partial_{x_n}^{\alpha_n}$, $\partial_{x_j} = \dfrac{\partial}{\partial x_j}$, $D_x^\alpha = D_{x_1}^{\alpha_1} \ldots D_{x_n}^{\alpha_n}$, $D_x = \dfrac{1}{i}\partial_x$, $D_{x_j} = \dfrac{1}{i}\partial_{x_j}$. If $X \subset \mathbb{R}^n$ is open, $k \in \mathbb{N}$, we let $C^k(X)$ denote the (Fréchet) space of k times continuously differentiable functions $X \to \mathbb{C}$. For $k = 0$ we get the space $C(X)$ of continuous complex-valued functions and we let $C^\infty(X) = \bigcap_{k \in \mathbb{N}} C^k(X)$ be the (Fréchet) space of infinitely (continuously) differentiable functions. If I is a subset of $\mathbb{R}$, $C^k(X; I)$ is the set of functions $\in C^k(X)$ taking their values in I. Recall that if $u \in C^k(X)$, the support of u, $\operatorname{supp} u$, is by definition the smallest closed subset L of X outside which u vanishes identically. For $k \in \mathbb{N} \cup \{\infty\}$ we let $C_0^k(X) = \{u \in C^k(X); \operatorname{supp} u \text{ is compact}\}$. If $M \subset \mathbb{R}^n$ is closed, we let $C^k(M)$ denote the space of restrictions to M of functions in $C^k(\mathbb{R}^n)$. For such functions we define the support as above as a closed subset of M, and we can then define $C_0^k(M)$ as the space of functions in $C^k(M)$ with bounded support.

Let $X \subset \mathbb{R}^n$ be an open set, $0 \leq \rho \leq 1$, $0 \leq \delta \leq 1$, $m \in \mathbb{R}$, $N \in \mathbb{N}\setminus\{0\}$. Then we have the following

Definition 1.1 $S^m_{\rho,\delta}(X \times \mathbb{R}^N)$ is the space of all $a \in C^\infty(X \times \mathbb{R}^N)$ such that for all compact $K \subset\subset X$ and all $\alpha \in \mathbb{N}^n$, $\beta \in \mathbb{N}^N$, there is a constant $C = C_{K,\alpha,\beta}(a)$ such that

$$|\partial_x^\alpha \partial_\theta^\beta a(x,\theta)| \leq C(1+|\theta|)^{m-\rho|\beta|+\delta|\alpha|}, \quad (x,\theta) \in K \times \mathbb{R}^N.$$

We say that $S^m_{\rho,\delta}$ is the space of symbols of order m and of type (ρ, δ).

It is easy to check that $S^m_{\rho,\delta}(X \times \mathbb{R}^N)$ is a Fréchet (vector) space with the seminorms :

$$P_{K,\alpha,\beta}(a) = \sup_{(x,\theta) \in K \times \mathbb{R}^N} \frac{|\partial_x^\alpha \partial_\theta^\beta a(x,\theta)|}{(1+|\theta|)^{m-\rho|\beta|+\delta|\alpha|}}$$

for K compact in X, $\alpha \in \mathbb{N}^n$, $\beta \in \mathbb{N}^N$. (A countable family of seminorms defining the topology is given by the $P_{K_j,\alpha,\beta}$, $j = 1, 2 \ldots$, $\alpha \in \mathbb{N}^n$, $\beta \in \mathbb{N}^N$, where $K_1, K_2, \ldots$ is an increasing sequence of compact subsets of X such that $X = \bigcup_{j=1}^\infty K_j$.) The operator $\partial_x^\alpha \partial_\theta^\beta$ is continuous from $S^m_{\rho,\delta}(X \times \mathbb{R}^N)$ to $S^{m-|\beta|\rho+|\alpha|\delta}_{\rho,\delta}(X \times \mathbb{R}^N)$.

If $m \leq m'$, $\delta \leq \delta'$, $\rho \geq \rho'$, then $S^m_{\rho,\delta} \subset S^{m'}_{\rho',\delta'}$. The space of symbols of order $-\infty$ is defined as

$$S^{-\infty}(X \times \mathbb{R}^N) = \{a \in C^\infty(X \times \mathbb{R}^N) \; ; \; \text{for every compact } K \subset X$$

and all $\alpha \in \mathbb{N}^n$, $\beta \in \mathbb{N}^N$, $M \in \mathbb{R}$, there exists $C = C_{K,\alpha,\beta,M}(a)$ such that

$$|\partial_x^\alpha \partial_\theta^\beta a(x,\theta)| \leq C(1+|\theta|)^{-M}, \; x \in K, \; \theta \in \mathbb{R}^N\}.$$

This space is also a Fréchet space (with the "best" constants as seminorms), and for every fixed $(\rho,\delta) \in [0,1] \times [0,1]$ we have

$$S^{-\infty}(X \times \mathbb{R}^N) = \bigcap_{m \in \mathbb{R}} S^m_{\rho,\delta}(X \times \mathbb{R}^N).$$

There is no point in introducing spaces $S^m_{\rho,\delta}(X \times \mathbb{R}^N)$ with $\rho > 1$ or with $\delta < 0$. For instance if $a \in S^m_{\rho,\delta}$ with $m < 0$ and $\rho > 1$, then applying $|\theta|\partial_{|\theta|} = \sum \theta_j \partial_{\theta_j}$ (working in polar coordinates) many times and then integrating, we see that $a \in S^{-\infty}$. Similar phenomena appear for $m \geq 0$, and also for $\delta < 0$.

Example 1.2 Let $a \in C^\infty(X \times \mathbb{R}^N)$ be positively homogeneous of degree m in the region $|\theta| \geq 1$; $a(x,\lambda\theta) = \lambda^m a(x,\theta)$, $\lambda \geq 1$, $|\theta| \geq 1$. Then $a \in S^m_{1,0}(X \times \mathbb{R}^N)$.

Example 1.3 $e^{ix\cdot\xi} \in S^0_{0,1}(\mathbb{R}^n \times \mathbb{R}^n)$. (Here $x\cdot\xi = \sum x_j \xi_j$, $x = (x_1,\ldots,x_n)$, $\xi = (\xi_1,\ldots,\xi_n)$.)

Example 1.4 Let $f \in C^\infty(X \times \mathbb{R}^N \; ; \; [0,\infty[)$ (i.e. f is smooth on $X \times \mathbb{R}^N$ and takes its values in $[0,\infty[)$, and let f be positively homogeneous of degree 1 for $|\theta| \geq 1$. Then $e^{-f} \in S^0_{\frac{1}{2},\frac{1}{2}}(X \times \mathbb{R}^N)$.

Verification : First, if $g \in C^2(\Omega)$ where $\Omega \subset \mathbb{R}^M$ is open and $g \geq 0$, then for every compact $K \subset \Omega$ we have $|g'(x)| \leq C \, g(x)^{1/2}$, $x \in K$, where $g'(x) = (\partial_{x_1} g(x),\ldots,\partial_{x_M} g(x))$. In fact, for $x \in K$ and $y \in \mathbb{R}^M$ sufficiently small, we have $0 \leq g(x+y) = g(x) + g'(x)\cdot y + C|y|^2$, so that $-g'(x)\cdot y \leq g(x) + C|y|^2$. We take $y = -r g'(x)$ with $r > 0$ sufficiently small depending on K and on g. Then we get $r|g'(x)|^2 \leq g(x) + Cr^2|g'(x)|^2$, so if $Cr \leq \frac{1}{2}$ we get the desired estimate with a new constant C. (In general we shall follow the convention that "C" denotes a new constant in every new formula.)

Applying this estimate to f and using the homogeneity, we find

$$|\theta|^{-\frac{1}{2}}|f'_x(x,\theta)| + |\theta|^{\frac{1}{2}}|f'_\theta(x,\theta)| \leq C_K |f(x,\theta)|^{\frac{1}{2}},$$

$$(x,\theta) \in K \times \mathbb{R}^N, \; |\theta| \geq 1, \; K \text{ compact } \subset X,$$

where $f'_x = (\partial_{x_1}f, \ldots, \partial_{x_n}f)$, $f'_\theta = (\partial_{\theta_1}f, \ldots, \partial_{\theta_N}f)$. Hence (using that $t^k e^{-t} \leq C_k$, $t \geq 0$, $k \geq 0$),

$$|f'_x|^k |f'_\xi|^\ell e^{-f} \leq C_{k,\ell,K}(1+|\theta|)^{\frac{k-\ell}{2}}, \quad (x,\theta) \in K \times \mathbb{R}^N,$$

and by induction we find that $\partial_x^\alpha \partial_\theta^\beta e^{-f}$ is a finite sum of terms of the type $a(x,\theta)(\partial_x f)^{\tilde{\alpha}}(\partial_\theta f)^{\tilde{\beta}} e^{-f}$ with $\tilde{\alpha} \in \mathbb{N}^n$, $\tilde{\beta} \in \mathbb{N}^N$, $|\tilde{\alpha}| \leq |\alpha|$, $|\tilde{\beta}| \leq |\beta|$, $a \in S_{1,0}^{(|\alpha|-|\tilde{\alpha}|)/2-(|\beta|-|\tilde{\beta}|)/2}$. From this it follows that $e^{-f} \in S_{\frac{1}{2},\frac{1}{2}}^0$.

Proposition 1.5 If $a \in S_{\rho,\delta}^{m_1}(X \times \mathbb{R}^N)$, $b \in S_{\rho,\delta}^{m_2}(X \times \mathbb{R}^N)$, then $ab \in S_{\rho,\delta}^{m_1+m_2}(X \times \mathbb{R}^N)$. More generally, the bilinear map

$$S_{\rho,\delta}^{m_1}(X \times \mathbb{R}^N) \times S_{\rho,\delta}^{m_2}(X \times \mathbb{R}^N) \ni (a,b) \mapsto ab \in S_{\rho,\delta}^{m_1+m_2}(X \times \mathbb{R}^N)$$

is continuous.

Outline of the proof: The first statement is immediate if we express $\partial_x^\alpha \partial_\theta^\beta (ab)$ by means of Leibniz' formula. The only additional work required for the second statement is to recall how to express the continuity of the bilinear map in terms of seminorms (see Exercise 1.5). □

Proposition 1.6 Let $(a_j)_{j=1}^\infty$ be a bounded sequence in $S_{\rho,\delta}^m(X \times \mathbb{R}^N)$ which converges at each point $(x,\theta) \in X \times \mathbb{R}^N$. Then the pointwise limit a belongs to $S_{\rho,\delta}^m(X \times \mathbb{R}^N)$ and for every $m' > m$ we have $a_j \to a$ in (the topology of) $S_{\rho,\delta}^{m'}(X \times \mathbb{R}^N)$.

Proof: For $f \in C^2([-\varepsilon,\varepsilon])$ (twice continuously differentiable on $[-\varepsilon,\varepsilon]$), $\varepsilon > 0$ we have:

$$(1.1) \quad |f'(0)| \leq C_\varepsilon(\|f\|_{L^\infty}^{\frac{1}{2}} \|f''\|_{L^\infty}^{\frac{1}{2}} + \|f\|_{L^\infty}), \quad \|f\|_{L^\infty} = \sup_{[-\varepsilon,\varepsilon]} |f(x)|.$$

(See Exercise 1.6.)

Applying this estimate to the various variables, we obtain by induction that $(a_j)_{j=1}^\infty$ is a Cauchy sequence in $C^k(X \times \mathbb{R}^N)$ for all $k \in \mathbb{N}$ and hence also for $k = \infty$. Consequently, the limit a belongs to $C^\infty(X \times \mathbb{R}^N)$ and $a_j \to a$ in $C^\infty(X \times \mathbb{R}^N)$. It is then clear that $a \in S_{\rho,\delta}^m$.

In order to prove the convergence in $S_{\rho,\delta}^{m'}$, we let K be compact $\subset X$, $\alpha \in \mathbb{N}^n$, $\beta \in \mathbb{N}^N$ and consider for $(x,\theta) \in K \times \mathbb{R}^N$:

$$k_j(x,\theta) = \frac{\partial_x^\alpha \partial_\theta^\beta (a_j - a)}{(1+|\theta|)^{m'-\rho|\beta|+\delta|\alpha|}} = \frac{1}{(1+|\theta|)^{m'-m}} \cdot \frac{\partial_x^\alpha \partial_\theta^\beta (a_j - a)}{(1+|\theta|)^{m-\rho|\beta|+\delta|\alpha|}}.$$

The last factor to the right is uniformly bounded with respect to j, x, θ, hence for every $\varepsilon > 0$, there is an $R_\varepsilon > 0$ such that $|k_j(x,\theta)| < \dfrac{\varepsilon}{2}$ for

$x \in K$, $(1+|\theta|) \geq R_\varepsilon$, $j = 1, 2 \ldots$. On the other hand $k_j(x,\theta) \to 0$, $j \to \infty$, uniformly on the compact set given by $x \in K$, $(1+|\theta|) \leq R_\varepsilon$. Hence $k_j(x,\theta) \to 0$ uniformly on $K \times \mathbb{R}^N$. It follows that $a_j - a \to 0$ in $S^{m'}_{\rho,\delta}(X \times \mathbb{R}^N)$. □

Proposition 1.7 *If $m' > m$, then $S^{-\infty}(X \times \mathbb{R}^N)$ is dense in $S^m_{\rho,\delta}(X \times \mathbb{R}^N)$ for the topology of $S^{m'}_{\rho,\delta}(X \times \mathbb{R}^N)$.*

Proof: Let $\chi \in C_0^\infty(\mathbb{R}^N)$, $\chi(\theta) = 1$ for $|\theta| \leq 1$, $\chi(\theta) = 0$ for $|\theta| \geq 2$. Then $\chi_j(\theta) = \chi\left(\dfrac{\theta}{j}\right)$, $j = 1, 2 \ldots$ is a bounded sequence in $S^0_{1,0}(X \times \mathbb{R}^N)$ (the symbols are actually independent of x). In fact, for $\alpha \in \mathbb{N}^N \setminus \{0\}$, $\partial_\theta^\alpha \chi_j(\theta) = j^{-|\alpha|} \chi^{(\alpha)}(\theta/j) = \mathcal{O}((1+|\theta|)^{-|\alpha|})$ uniformly in j, θ, since $j \leq |\theta| \leq 2j$ on the support of $\chi^{(\alpha)}(\theta/j)$. (Here $\chi^{(\alpha)} = \partial_\theta^\alpha \chi$.) If $a \in S^m_{\rho,\delta}(X \times \mathbb{R}^N)$ it suffices to put $a_j(x,\theta) = \chi\left(\dfrac{\theta}{j}\right) a(x,\theta)$ and apply Propositions 1.5, 1.6 as well as the inclusion $S^0_{1,0} \subset S^0_{\rho,\delta}$. □

We next study asymptotic sums of symbols.

Proposition 1.8 *Let $a_j \in S^{m_j}_{\rho,\delta}(X \times \mathbb{R}^N)$, $j = 0, 1, 2 \ldots$ with $m_j \searrow -\infty$, $j \to \infty$. Then there exists $a \in S^{m_0}_{\rho,\delta}(X \times \mathbb{R}^N)$ unique modulo (i.e. up to some element in) $S^{-\infty}(X \times \mathbb{R}^N)$, such that $a - \sum_{0 \leq j < k} a_j \in S^{m_k}_{\rho,\delta}$ for $k = 0, 1, 2, \ldots$.*

Proof: The uniqueness modulo $S^{-\infty}(X \times \mathbb{R}^N)$ follows from: $\bigcap\limits_{k=0}^{\infty} S^{m_k}_{\rho,\delta} = S^{-\infty}$. As for the existence, we may first assume that the sequence (m_j) is strictly decreasing, since we could otherwise regroup the terms with the same value of m_j. For each space $S^{m_j}_{\rho,\delta}(X \times \mathbb{R}^N)$, let $P_{j,0}, P_{j,1}, P_{j,2}, \ldots$ be a sequence of seminorms defining the topology on $S^{m_j}_{\rho,\delta}$. According to Proposition 1.7, for each j we can find $b_j \in S^{-\infty}$ such that $P_{\nu,\mu}(a_j - b_j) \leq 2^{-j}$ for $0 \leq \nu$, $\mu \leq j - 1$. Then $\sum\limits_{j \geq k}(a_j - b_j)$ converges in $S^{m_k}_{\rho,\delta}$ for each k, and if we put $a = \sum\limits_0^\infty (a_j - b_j) \in S^{m_0}_{\rho,\delta}$ then

$$a - \sum_{j<k} a_j = -\sum_{j<k} b_j + \sum_k^\infty (a_j - b_j) \in S^{m_k}_{\rho,\delta}.$$

□

This proof uses the standard Cantor diagonalization procedure and is close to the Borel theorem which states that for any family $a_\alpha \in \mathbb{C}$, $\alpha \in \mathbb{N}^n$, there exists $f \in C^\infty(\mathbb{R}^n)$ with $f^{(\alpha)}(0) = \alpha! a_\alpha$, $\alpha \in \mathbb{N}^n$, so that the (formal) Taylor–Maclaurin series of f at 0 is $\sum_{\alpha \in \mathbb{N}^n} a_\alpha x^\alpha$. See Exercise 1.7.

If a and a_j have the properties of the proposition, we write $a \sim \sum_j^\infty a_j$ and we call a an (and sometimes "the") asymptotic sum of the a_j. In practice we often need the following result :

Proposition 1.9 Let $a_j \in S_{\rho,\delta}^{m_j}(X \times \mathbb{R}^N)$, $j = 0, 1, 2\ldots$, with $m_j \searrow -\infty$ and let $a \in C^\infty(X \times \mathbb{R}^N)$ have the properties
1) For all compact $K \subset X$ and $\alpha \in \mathbb{N}^n$, $\beta \in \mathbb{N}^N$, there exists $C_{\alpha,\beta,K}$, $M_{\alpha,\beta} > 0$ such that

$$|\partial_x^\alpha \partial_\theta^\beta a(x,\theta)| \le C_{\alpha,\beta,K}(1+|\theta|)^{M_{\alpha,\beta}}, \quad (x,\theta) \in K \times \mathbb{R}^N.$$

2) There exists a sequence $m_k' \to -\infty$, $k \to \infty$ such that for all compact $K \subset X$ and $k \in \mathbb{N}$, there exists $C_{K,k} > 0$ such that

$$|a(x,\theta) - \sum_0^{k-1} a_j(x,\theta)| \le C_{K,k}(1+|\theta|)^{m_k'}, \quad (x,\theta) \in K \times \mathbb{R}^N.$$

Then $a \sim \sum_0^\infty a_j$.

Proof: Let $a' \in S_{\rho,\delta}^{m_0}$ be an asymptotic sum, $a' \sim \sum_0^\infty a_j$. Then $b = a - a'$ has the property 1), and according to 2) we know that for every compact $K \subset X$ and $M \in \mathbb{N}$, there exists $C_{K,M} > 0$ such that $|b(x,\theta)| \le C_{K,M}(1+|\theta|)^{-M}$ on $K \times \mathbb{R}^N$.

Using (1.1) we find for every sufficiently small $\varepsilon > 0$ and for $(\alpha,\beta) \in \mathbb{N}^{n+N}$, $|(\alpha,\beta)| = 1$:

$$|\partial_x^\alpha \partial_\theta^\beta b(x,\theta)| \le C_\varepsilon\left((\sup_B |b|)^{\frac{1}{2}}(\sup_B |(\partial_x^\alpha \partial_\theta^\beta)^2 b|)^{\frac{1}{2}} + \sup_B |b|\right),$$

where $B = B((x,\theta),\varepsilon)$ denotes the open Euclidean ball in $\mathbb{R}^{n+N}$ of center (x,θ) and of radius ε. We conclude that

$$|\partial_x^\alpha \partial_\theta^\beta b(x,\theta)| \le \tilde{C}_{K,M}(1+|\theta|)^{-M} \quad \text{on} \quad K \times \mathbb{R}^N,$$

for every compact $K \subset X$ and every $M \in \mathbb{N}$. Iterating this argument we get $b \in S^{-\infty}$ and hence $a \sim \sum_0^\infty a_j$. $\square$

From now on, we assume that $0 < \rho \le 1$, $0 \le \delta < 1$.

Definition 1.10 A function $\varphi = \varphi(x,\theta) \in C^\infty(X \times \dot{\mathbb{R}}^N)$ is called a phase function if for all $(x,\theta) \in X \times \dot{\mathbb{R}}^N$:

1) $\operatorname{Im}\varphi(x,\theta) \geq 0$,

2) $\varphi(x,\lambda\theta) = \lambda\varphi(x,\theta)$ for all $\lambda > 0$,

3) $d\varphi \neq 0$.

Here $d\varphi = \sum_1^n \frac{\partial \varphi}{\partial x_j} dx_j + \sum_1^N \frac{\partial \varphi}{\partial \theta_k} d\theta_k$ so 3) means that at every point some derivative $\frac{\partial \varphi}{\partial x_j}$ or $\frac{\partial \varphi}{\partial \theta_k}$ is non-vanishing.

If φ is a phase function and $a \in S^m_{\rho,\delta}(X \times \mathbb{R}^N)$, $m+k < -N$, $k \in \mathbb{N}$ then

$$(1.2) \qquad I(a,\varphi)(x) \stackrel{\text{def}}{=} \int e^{i\varphi(x,\theta)} a(x,\theta)\, d\theta \in C^k(X),$$

and the map $S^m_{\rho,\delta} \ni a \mapsto I(a,\varphi) \in C^k(X)$ is continuous.

We let $\mathcal{D}'(X)$ be the space of (Schwartz) distributions on X (it is the dual space of $\mathcal{D}(X) = C_0^\infty(X)$,) and $\mathcal{D}'^{(k)}(X)$ be the subspace of distributions of order $\leq k$ (it is the dual space of $C_0^k(X)$). The natural duality between distributions and test functions will be expressed by $\langle,\rangle$ and sometimes we write formally $\langle u, \varphi \rangle = \int u(x)\varphi(x)\, dx$, $u \in \mathcal{D}'(X)$, $\varphi \in C_0^\infty(X)$. We shall only use the weak topology on $\mathcal{D}'(X)$ and on $\mathcal{E}'(X) = \{u \in \mathcal{D}'(X); \operatorname{supp} u$ is compact$\}$.

Theorem 1.11 *Let $\varphi(x,\theta)$ be a phase function on $X \times \dot{\mathbb{R}}^N$ and let $0 < \rho \leq 1, 0 \leq \delta < 1$. Then there is a unique way of defining $I(a,\varphi) \in \mathcal{D}'(X)$ for $a \in S^\infty_{\rho,\delta} \stackrel{\text{def}}{=} \bigcup_{m \in \mathbb{R}} S^m_{\rho,\delta}$ such that $I(a,\varphi)$ is defined by (1.2) when $a \in S^m_{\rho,\delta}$, $m < -N$ and such that for every $m \in \mathbb{R}$, the map $S^m_{\rho,\delta} \ni a \mapsto I(a,\varphi) \in \mathcal{D}'(X)$ is continuous.*

Moreover, if $k \in \mathbb{N}$ and $m - k\min(\rho, 1-\delta) < -N$, then $S^m_{\rho,\delta} \ni a \mapsto I(a,\varphi) \in \mathcal{D}'^{(k)}(X)$ is continuous.

Proof: The uniqueness is obvious since $S^{-\infty}$ is dense in $S^m_{\rho,\delta}$ for the topology of $S^{m'}_{\rho,\delta}$ if $m' > m$. For the existence, we shall define $I(a,\varphi)$ with the help of formal integrations by parts, using the following result:

Lemma 1.12 *There exist $a_j \in S^0_{1,0}(X \times \mathbb{R}^N)$, $b_j, c \in S^{-1}_{1,0}(X \times \mathbb{R}^N)$ such that the differential operator*

$$L = \sum a_j(x,\theta) \frac{\partial}{\partial \theta_j} + \sum b_j(x,\theta) \frac{\partial}{\partial x_j} + c(x,\theta)$$

satisfies : ${}^tL(e^{i\varphi}) = e^{i\varphi}$. Here ${}^tL = -\sum \frac{\partial}{\partial \theta_j} \circ a_j - \sum \frac{\partial}{\partial x_j} \circ b_j + c$ is the real transpose of L satisfying $\iint L(f)g\, dx\, d\theta = \iint f\, {}^tL(g)\, dx\, d\theta$, $f, g \in C_0^\infty(X \times \mathbb{R}^N)$.

Proof of the lemma: The function
$$\Phi(x,\theta) = \sum_1^n \overline{\frac{\partial \varphi}{\partial x_j}} \frac{\partial \varphi}{\partial x_j} + |\theta|^2 \sum_1^N \overline{\frac{\partial \varphi}{\partial \theta_j}} \frac{\partial \varphi}{\partial \theta_j}$$
is $\neq 0$ for $|\theta| \neq 0$ and positively homogeneous of degree 2 : $\Phi(x, \lambda\theta) = \lambda^2 \Phi(x,\theta)$, $\theta \neq 0$, $\lambda > 0$. Let $\chi(\theta) \in C_0^\infty(\mathbb{R}^N)$ be equal to 1 in a neighborhood of 0 and put

$$\begin{aligned}{}^tL &= \frac{1-\chi(\theta)}{i\,\Phi(x,\theta)} \left(\sum_1^N |\theta|^2 \overline{\frac{\partial \varphi}{\partial \theta_j}} \frac{\partial}{\partial \theta_j} + \sum_1^n \overline{\frac{\partial \varphi}{\partial x_j}} \frac{\partial}{\partial x_j} \right) + \chi(\theta) \\ &= \sum a_j' \frac{\partial}{\partial \theta_j} + \sum b_j' \frac{\partial}{\partial x_j} + c'\end{aligned}$$

with $a_j' \in S_{1,0}^0$, $b_j' \in S_{1,0}^{-1}$, $c' \in S^{-\infty}$. Then ${}^tL(e^{i\varphi}) = e^{i\varphi}$, and $L = {}^t({}^tL)$ will have an expression as in the lemma. □

For $u \in C_0^\infty(X)$, $a \in S^{-\infty}(X \times \mathbb{R}^N)$, we have

(1.3) $\langle I(a,\varphi), u \rangle = \iint e^{i\varphi(x,\theta)} a(x,\theta) u(x) dx\, d\theta$
$= \iint ({}^tL)^k(e^{i\varphi}) au\, dx\, d\theta = \iint e^{i\varphi(x,\theta)} L^k(a(x,\theta)u(x))\, dx\, d\theta.$

If $a \in S_{\rho,\delta}^m$, $u \in C_0^\infty(X)$, then $L^k(au) \in S_{\rho,\delta}^{m-kt}$, where we have put $t = \min(\rho, 1-\delta)$. More precisely, we have the continuous map

(1.4) $S_{\rho,\delta}^m(X \times \mathbb{R}^N) \times C_0^\infty(X) \ni (a,u) \mapsto L^k(au) \in S_{\rho,\delta}^{m-kt}(X \times \mathbb{R}^N).$

For every compact $K \subset \Omega$,
$$\sup_{K \times \mathbb{R}^N} |L^k(au)|(1+|\theta|)^{-m+kt} \leq f_{k,K}(a) \sum_{|\alpha| \leq k} \sup_K |\partial^\alpha u(x)|,$$
where $f_k(a)$ is a suitable seminorm on $S_{\rho,\delta}^m$.

For $a \in S_{\rho,\delta}^m$ we choose $k \in \mathbb{N}$ with $m - kt < -N$ and we put

(1.5) $\langle I_k(a,\varphi), u \rangle = \iint e^{i\varphi} L^k(au)\, dx\, d\theta.$

Then (1.4) shows that $I_k(a,\varphi) \in \mathcal{D}'(X)$ and the estimate after (1.4) shows that $I_k(a,\varphi) \in \mathcal{D}'^{(k)}$ and that the map $S_{\rho,\delta}^m \ni a \mapsto I_k(a,\varphi) \in \mathcal{D}'^{(k)}(X)$ is continuous.

Moreover $I_k(a,\varphi) = I(a,\varphi)$ when $a \in S^{-\infty}$, and by a density argument we have $I_{k'}(a,\varphi) = I_{k''}(a,\varphi)$ for $a \in S^m_{\rho,\delta}$ if $m - k't < -N$, $m - k''t < -N$. We can then define $I(a,\varphi) = I_k(a,\varphi)$ for any k with $m - kt < -N$. □

$I(a,\varphi)$ is called an oscillatory integral or a Fourier integral distribution. Even if the order of a is large we shall write formally

$$I(a,\varphi)(x) = \int e^{i\varphi(x,\theta)} a(x,\theta)\, d\theta,$$

$$\langle I(a,\varphi), u \rangle = \int\int e^{i\varphi(x,\theta)} a(x,\theta) u(x)\, dx\, d\theta$$

$\Big($ If $\chi \in \mathcal{S}(\mathbb{R}^N)$, $\chi(0) = 1$, then

$$I(a,\varphi) = \lim_{\substack{\varepsilon \to 0 \\ \mathcal{D}'(X)}} \int e^{i\varphi(x,\theta)} \chi(\varepsilon\theta) a(x,\theta)\, d\theta.\Big)$$

We also notice for later use that if $L = \sum a_j(x,\theta)\dfrac{\partial}{\partial \theta_j} + \sum b_j(x,\theta)\dfrac{\partial}{\partial x_j} + c(x,\theta)$ with $a_j \in S^0_{\rho,\delta}$, $b_j, c \in S^{-1}_{\rho,\delta}$, $a \in S^m_{\rho,\delta}$, $u \in C^\infty_0(X)$ then

$$\int\int {}^tL(e^{i\varphi})a(x,\theta)u(x)\,dx\,d\theta = \int\int e^{i\varphi(x,\theta)} L(a(x,\theta)u(x))\,dx\,d\theta.$$

Definition 1.13 If $\varphi \in C^\infty(X \times \dot{\mathbb{R}}^N)$ is a phase function we call $C_\varphi = \{(x,\theta) \in X \times \dot{\mathbb{R}}^N\,;\, \varphi'_\theta = 0\}$ the critical set of φ. $\Big($Here $\varphi'_\theta = \Big(\dfrac{\partial\varphi}{\partial\theta_1}, \ldots, \dfrac{\partial\varphi}{\partial\theta_N}\Big).\Big)$

It is the behavior of a, φ near C_φ which determines the singularities of $I(a,\varphi)$:

Lemma 1.14 If $a \in S^m_{\rho,\delta}(X \times \mathbb{R}^N)$ vanishes in a conical neighborhood of C_φ, then $I(a,\varphi) \in C^\infty(X)$.

Proof: Repeating the proof of Lemma 1.12, we can construct $L\Big(x, \theta, \dfrac{\partial}{\partial \theta}\Big) = \sum a_j(x,\theta) \dfrac{\partial}{\partial \theta_j} + c(x,\theta)$, $a_j \in S^0_{1,0}$, $c \in S^{-1}_{1,0}$ such that ${}^tL(e^{i\varphi}) = (1+b)\,e^{i\varphi}$, where supp $b \cap$ supp $a = \emptyset$. As in the proof of Theorem 1.11, we get $I(a,\varphi) = I(L^k(a),\varphi)$. On the other hand $L^k(a) \in S^{m-kt}_{\rho,\delta}$ so $I(a,\varphi) \in C^k(X)$ for every $k \in \mathbb{N}$, hence $I(a,\varphi) \in C^\infty(X)$. □

Corollary 1.15 Let Π denote the natural projection: $\Pi : X \times \mathbb{R}^N \ni (x,\theta) \mapsto x \in X$. Then sing supp $I(a,\varphi) \subset \Pi(C_\varphi)$.

We recall that the singular support, sing supp u, of a distribution $u \in \mathcal{D}'(X)$ is the smallest closed subset L of X such that $u_{|X \setminus L}$ is of class C^∞.

Example 1.16 a) The Dirac measure (at zero) in $\mathbb{R}^n$ is given by $\delta(x) = \int e^{ix\xi} \frac{d\xi}{(2\pi)^n}$ and hence $D^\alpha \delta = \frac{\delta^{(\alpha)}}{i^{|\alpha|}}(x) = \int e^{ix\cdot\xi} \xi^\alpha \frac{d\xi}{(2\pi)^n}$. We have sing supp $(D^\alpha \delta) = \{0\}$.

b) If $f(x) \in C^\infty(X)$, $\operatorname{Im} f \geq 0$, $df(x) \neq 0$ whenever $f(x) = 0$, then

$$\frac{1}{(f(x)+i0)} = \lim_{\varepsilon \searrow 0} \frac{1}{(f(x)+i\varepsilon)} = \frac{1}{i} \int_0^\infty e^{if(x)\tau}\, d\tau.$$

(To give a sense to the last integral we introduce a partition of unity $1 = \chi(\tau) + (1-\chi(\tau))$, where $\chi \in C_0^\infty(\mathbb{R})$, $\chi(\tau) = 1$ in a neighborhood of 0.) See Exercise 1.9.

We end this chapter by considering operators. Let $X \subset \mathbb{R}^{n_X}$, $Y \subset \mathbb{R}^{n_Y}$ be open sets. We recall that the Schwartz kernel theorem states that there is a bijection between the set of distributions $K \in \mathcal{D}'(X \times Y)$ and the set of continuous linear operators $A : C_0^\infty(Y) \to \mathcal{D}'(X)$. The correspondence is given by

$$(1.6) \qquad \langle Au, v \rangle_X = \langle K, v \otimes u \rangle_{X \times Y}, \quad u \in C_0^\infty(Y),\ v \in C_0^\infty(X),$$

where the subscripts indicate that we use the duality brackets for $\mathcal{D}'(X) \times C_0^\infty(X)$ and $\mathcal{D}'(X \times Y) \times C_0^\infty(X \times Y)$ respectively. Also, $(v \otimes u)(x,y) = v(x)\,u(y)$. We call K the distribution kernel of A, we sometimes write $K = K_A$ and sometimes we even use the same symbol to denote an operator and its distribution kernel.

If φ is a phase function on $X \times Y \times \dot{\mathbb{R}}^N$ and $a \in S^m_{\rho,\delta}(X \times Y \times \mathbb{R}^N)$ (where the former "base variables" x are now replaced by (x,y)), then $K = I(a,\varphi) \in \mathcal{D}'(X \times Y)$ is the distribution kernel of an operator A which we can formally express as

$$(1.7) \qquad Au(x) = \int\int e^{i\varphi(x,y,\theta)} a(x,y,\theta)\, u(y)\, dy\, d\theta, \quad u \in C_0^\infty(Y).$$

Such an operator is called a Fourier integral operator (FIO). For example, if $X = Y$ and $\varphi = \varphi(x,y,\xi) = (x-y)\xi$ then by definition, A is a pseudodifferential operator.

Theorem 1.17 Let $\varphi \in C^\infty(X \times Y \times \dot{\mathbb{R}}^N)$ and $a \in S^m_{\rho,\delta}(X \times Y \times \mathbb{R}^N)$. Let A be given by (1.7) (so that $K_A = I(a,\varphi)$).

(A) If for every $x \in X$, $(y,\theta) \mapsto \varphi(x,y,\theta)$ is a phase function, then A is continuous $C_0^\infty(Y) \to C^\infty(X)$.

(B) If for every $y \in Y$, $(x, \theta) \mapsto \varphi(x, y, \theta)$ is a phase function, then A has a (unique) continuous extension $\mathcal{E}'(Y) \to \mathcal{D}'(X)$.

Proof: (A) Since $d_{(y,\theta)}\varphi \neq 0$, we can find

$$L = L\left(x, y, \theta, \frac{\partial}{\partial y}, \frac{\partial}{\partial \theta}\right) \in S^0_{1,0}\frac{\partial}{\partial \theta} \oplus S^{-1}_{1,0}\frac{\partial}{\partial y} \oplus S^{-1}_{1,0}$$

such that ${}^tL(e^{i\varphi}) = e^{i\varphi}$ and then for $u \in C_0^\infty(Y)$, $v \in C_0^\infty(X)$ and $k \in \mathbb{N}$ with $m - kt < -N$:

$$\langle Au, v\rangle = \langle K_A, v \otimes u\rangle = \int\int\int e^{i\varphi(x,y,\theta)} v(x) L^k(a(x,y,\theta)u(y))\, dx\, dy\, d\theta.$$

Hence $Au(x)$ is the C^∞ function given by

(1.8) $$Au(x) = \int\int e^{i\varphi(x,y,\theta)} L^k(a(x,y,\theta)\, u(y))\, dy\, d\theta.$$

The continuity $C_0^\infty(Y) \to C^\infty(X)$ also follows from this formula.

(B) We recall that the (real) transpose ${}^tA : C_0^\infty(X) \to \mathcal{D}'(Y)$ is defined (in view of Schwartz' kernel theorem) by

(1.9) $$\langle u, {}^tAv\rangle = \langle Au, v\rangle = \langle K, v \otimes u\rangle, \quad u \in C_0^\infty(Y),\ v \in C_0^\infty(X).$$

In other words, the distribution kernel of ${}^tA$ is obtained from that of A by exchanging the roles of x and y. According to (A), the hypothesis of (B) implies that ${}^tA : C_0^\infty(X) \to C^\infty(Y)$ is continuous. Then we define $Au \in \mathcal{D}'(X)$ for $u \in \mathcal{E}'(Y)$ by

$$\langle u, {}^tAv\rangle = \langle Au, v\rangle, \quad v \in C_0^\infty(X).$$

This gives a continuous extension and the uniqueness of our extension follows from the fact that $C_0^\infty(Y)$ is dense in $\mathcal{E}'(Y)$. □

Example 1.18 Pseudodifferential operators are continuous $C_0^\infty \to C^\infty$ and (have unique continuous extensions) $\mathcal{E}' \to \mathcal{D}'$.

Remark 1.19 If $\Gamma \subset X \times \dot{\mathbb{R}}^N$ is an open cone we have an obvious notion of phase functions $\varphi \in C^\infty(\Gamma)$. If φ is such a phase function and $a \in S^m_{\rho,\delta}(X \times \mathbb{R}^N)$ has the property that $\operatorname{supp} a \subset \Gamma'$, where $\Gamma' \subset \Gamma$ is a cone which is closed in $X \times \dot{\mathbb{R}}^N$, then we can still define $I(a, \varphi) \in \mathcal{D}'(X)$, and all the results of this chapter can naturally be extended to this case.

Exercises

Exercise 1.1

a) Give an example of a function $\varphi \in C^\infty(\mathbb{R}; [0, +\infty[)$ such that :
$$\begin{cases} \varphi(x) = 0 & x \leq 0 \\ \varphi(x) > 0 & x > 0. \end{cases}$$

b) Let $B = B(0,1)$ be the open unit ball in $\mathbb{R}^n$, $B = \{x \in \mathbb{R}^n; |x| < 1\}$. Find $\chi \in C_0^\infty(\mathbb{R}^n; [0, +\infty[)$ with $\operatorname{supp} \chi \subset B$ such that $\int_{\mathbb{R}^n} \chi(x) dx = 1$.

Exercise 1.2

Let K be a closed subset of $\mathbb{R}^n$. Find $\psi \in C^\infty(\mathbb{R}^n; [0,1])$ such that $\psi(x) = 1$ in a neighborhood of K and $\operatorname{supp} \psi \subset K + B(0, 2\varepsilon)$, $\varepsilon > 0$.

Exercise 1.3

Let X be an open subset of $\mathbb{R}^n$.

a) Find a sequence of compact sets K_j, $j \in \mathbb{N}$, such that $K_j \subset \mathring{K}_{j+1}$, $X = \bigcup_j K_j$, and such that for every compact set $K \subset X$ there exists j_0 with $K \subset \mathring{K}_j$, $j \geq j_0$.

b) Find a sequence (φ_j), $\varphi_j \in C_0^\infty(X)$ such that $\#\{j \,;\, K \cap \operatorname{supp} \varphi_j \neq \emptyset\}$ is finite for every compact set $K \subset X$ and such that $1 = \sum \varphi_j(x)$. Here $\#A$ is by definition the number of elements in A.

Such a sequence is called a locally finite partition of unity.

c) Suppose $X = \bigcup_{i \in I} A_i$ with A_i open and relatively compact in X.

Find a sequence (φ_j) as in b) such that for every j $\operatorname{supp} \varphi_j \subset A_i$ for some $i \in I$.

Exercise 1.4

Let X be an open subset of $\mathbb{R}^n$. Which of the spaces $C^\infty(X)$, $C_0^\infty(X)$ are Fréchet spaces ?

Let K be a compact subset of $\mathbb{R}^n$. Is $C^\infty(X) \cap \mathcal{E}'(K)$ a Fréchet space ? Here by definition, $\mathcal{E}'(K) = \{u \in \mathcal{E}'(\mathbb{R}^n)\,;\, \operatorname{supp} u \subset K\}$.

Exercise 1.5

Let F_1, F_2, F_3 be Fréchet spaces.

a) Recall how to express that a linear map $u : F_1 \to F_2$ is continuous, in terms of families of seminorms on F_1, F_2.

b) Let $B : F_1 \times F_2 \to F_3$ be a bilinear map. Express that B is continuous, in terms of seminorms, starting by discussing continuity at $(0,0)$.

Exercise 1.6

a) Let $f \in C^2([0,\varepsilon])$, $\varepsilon > 0$. Assume $f(0) \geq 0$, $f'(0) \geq 0$. Try to get a lower bound on $\sup_{0 \leq t \leq \varepsilon} f(t)$ using also $\|f''\|_{L^\infty}$.

b) Show inequality (1.1) : for $f \in C^2([-\varepsilon, \varepsilon])$ $\varepsilon > 0$

$$|f'(0)| \leq C_\varepsilon (\|f\|_{L^\infty}^{1/2} \|f''\|_{L^\infty}^{1/2} + \|f\|_{L^\infty}).$$

Exercise 1.7

Prove Borel's theorem : for every sequence $(a_\alpha)_{\alpha \in \mathbb{N}^n}$, $a_\alpha \in \mathbb{C}$, there exists $u \in C^\infty(\mathbb{R}; \mathbb{C})$ such that $a_\alpha = \dfrac{1}{\alpha!} \partial_x^\alpha u(0)$, for every α.

a) Consider $\varphi \in C^\infty(\mathbb{R}^n; [0,1])$ such that $\varphi(x) = 1$ if $\|x\| \leq \dfrac{1}{2}$ and $\varphi(x) = 0$ if $\|x\| \geq 1$ (see Exercise 1.2). Let

$$u_N(x, \lambda) = \varphi(\lambda x) \sum_{|\alpha|=N} a_\alpha x^\alpha \quad \text{for} \quad \lambda > 0.$$

Calculate $\partial_x^\beta u_N(0)$.

Show that $\|u_N(x, \lambda)\|_{C^{N-1}} \leq 2^{-N}$ if $\lambda \geq \lambda_N$, provided that $\lambda_N > 0$ is sufficiently large. Here we put $\|u\|_{C^k} = \sum_{|\alpha| \leq k} \sup_{x \in \mathbb{R}^n} |\partial^\alpha u(x)|$.

b) Show that $u(x) = \sum_N u_N(x, \lambda_N)$ is a solution of the problem.

Exercise 1.8

Let $\xi = (\xi', \xi_n) \in \mathbb{R}^n$, $\xi'^2 = \sum_{j=1}^{n-1} \xi_j^2$, $\xi^2 = \xi'^2 + \xi_n^2$. To which symbol spaces do the following symbols belong ?

a) $(\xi'^2 + i\xi_n)^{-1}$ b) $(\xi^2 + 1)^{-1}$ c) $(\xi'^2 + 1)^{-1}$

Exercise 1.9

a) Show that the Dirac measure at zero in $\mathbb{R}^n$ is given by $\delta = \displaystyle\int e^{ix\xi} \dfrac{d\xi}{(2\pi)^n}$ in the sense of oscillatory integrals.

b) Write $D^\alpha \delta = i^{-|\alpha|} \partial_x^\alpha \delta$ as an oscillatory integral and show that $\operatorname{sing\,supp}(D^\alpha \delta) = \{0\}$.

c) If $f \in C^\infty(X)$, $\operatorname{Im} f \geq 0$, $X \subset \mathbb{R}^n$ open, $\varepsilon > 0$, show that

$$\frac{1}{f(x)+i\varepsilon} = \frac{1}{i}\int_0^\infty e^{i(f(x)+i\varepsilon)\tau}\,d\tau.$$

d) Assume moreover that $df(x) \neq 0$ whenever $f(x) = 0$. Show that, with convergence in $\mathcal{D}'(X)$:

$$\frac{1}{f(x)+i0} \stackrel{\text{def}}{=} \lim_{\varepsilon\to 0} \frac{1}{f(x)+i\varepsilon} = \frac{1}{i}\int_0^\infty e^{if(x)\tau}\,d\tau,$$

and give a sense to the last integral by introducing $1 = \chi(\tau) + (1-\chi(\tau))$ with $\chi(\cdot) \in C_0^\infty(\mathbb{R})$, $\chi(\tau) = 1$ near 0.

e) What is $\operatorname{sing\,supp}\left(\dfrac{1}{f(x)+i0}\right)$? (Give two explanations.)

f) For $n = 1$ show that

$$\delta = \frac{1}{2i\pi}\left(\frac{1}{x-i0} - \frac{1}{x+i0}\right).$$

Exercise 1.10

Let $n > 2$ and $\Delta = \sum_{j=1}^n (\partial_{x_j})^2$.

a) Check that $\dfrac{1}{|x|^{n-2}} \in L^1_{\text{loc}}(\mathbb{R}^n)$ and calculate

$$\left\langle \Delta\left(\frac{1}{|x|^{n-2}}\right), \varphi \right\rangle, \quad \varphi \in C_0^\infty(\mathbb{R}^n).$$

b) Show that if Ω is an open subset of $\mathbb{R}^n$ and $u, v \in \mathcal{D}'(\Omega)$ satisfy $\Delta u = v$ then $\operatorname{sing\,supp} u = \operatorname{sing\,supp} v$.

Exercise 1.11

Consider the continuous linear map $A : C_0^\infty(Y) \to \mathcal{D}'(X)$, with distribution kernel $K_A \in \mathcal{D}'(X \times Y)$.

Show that i) and ii) are equivalent :
i) A can be extended to a continuous operator $\mathcal{E}'(Y) \to C^\infty(X)$.
ii) $K_A \in C^\infty(X \times Y)$.
(To prove i) $\implies$ ii), show that if $\varphi \in C_0^\infty(X)$, $\psi \in C_0^\infty(Y)$, then the Fourier transform of $\varphi(x)\psi(y)K_A(x,y)$ is rapidly decreasing.)

Exercise 1.12

Let $x = (x', x'') \in \mathbb{R}^{n-d} \times \mathbb{R}^d = \mathbb{R}^n$. For $u \in C_0^\infty(\mathbb{R}^n)$ let

$$Vu(x') = \int u(x', x'') \, dx''.$$

Write V as a Fourier integral operator and show that V can be extended to a continuous operator $\mathcal{E}'(\mathbb{R}^n) \to \mathcal{E}'(\mathbb{R}^{n-d})$.

Exercise 1.13

Let $f : X \to Y$ be a C^∞ diffeomorphism, X and Y open subsets of $\mathbb{R}^n$. Let $T : C_0^\infty(Y) \to C^\infty(X)$ be defined by

$$Tu(x) = u(f(x)).$$

Show that T is a Fourier integral operator.

Exercise 1.14

Give a more precise version of Theorem 1.17 in terms of the spaces C_0^k, C^k, $\mathcal{E}'^{(k)}$, $\mathcal{D}'^{(k)}$.

Notes

For general distribution theory see for instance L. Schwartz [Sch].

The symbol classes $S_{\rho,\delta}^m$ and oscillatory integral $I(a,\varphi)$ were introduced by L. Hörmander. (See [Hö2].) More special symbol spaces were previously used by Kohn–Nirenberg [KN] and more general ones are now also frequently used ; see [BF], [B], [Hö3,4], [R].

2 The method of stationary phase

Let $X \subset \mathbb{R}^n$ be an open set, $\varphi \in C^\infty(X;\mathbb{R})$ (i.e. a real-valued smooth function) such that $d\varphi \neq 0$ everywhere. If $u \in C_0^\infty(X)$, then the integral

$$(2.1) \qquad I(\lambda) = \int e^{i\lambda\varphi(x)} u(x)\, dx$$

is rapidly decreasing when $\lambda \to \infty$. This can be seen by using repeated integration by parts, using for instance the operator ${}^t L = \dfrac{1}{i\lambda|\varphi'|^2} \sum \dfrac{\partial \varphi}{\partial x_j} \dfrac{\partial}{\partial x_j}$. More precisely we obtain :

> For every compact $K \subset X$ and every $N \in \mathbb{N}$, there is a constant $C = C_{K,\varphi,N}$ such that

$$(2.2) \qquad |I(\lambda)| \leq C \Big(\sum_{|\alpha|\leq N} \sup |\partial^\alpha u(x)| \Big) \lambda^{-N},$$

$\lambda \geq 1$, $u \in C_0^\infty(X)$, supp $u \subset K$.

Hence in general, if $\varphi \in C^\infty(X;\mathbb{R})$, $u \in C_0^\infty(X)$, the asymptotic behavior of $I(\lambda)$ when $\lambda \to \infty$ is determined by φ, u in a neighborhood of the set of critical points of φ. (A point $x_0 \in X$ is a critical point of φ if $\varphi'(x_0) = 0$.) The most important (and easiest) case is that of a non-degenerate critical point. We say that $x_0 \in X$ is a non-degenerate critical point of φ if $\varphi'(x_0) = 0$ and $\det(\varphi''(x_0)) \neq 0$, where $\varphi''(x) = \Big(\dfrac{\partial^2 \varphi(x)}{\partial x_j \partial x_k}\Big)_{1\leq j,k \leq n}$ is the Hessian of φ at x. Since $\varphi''(x_0)$ is the differential of the map $X \ni x \mapsto \varphi'(x) \in \mathbb{R}^n$ at x_0, it follows that x_0 is an isolated critical point if it is a non-degenerate one : $\varphi'(x) \neq 0$ if $|x - x_0| > 0$ is sufficiently small. The Morse lemma gives local coordinates near a non-degenerate critical point x_0 of φ for which $\varphi(x) - \varphi(x_0)$ becomes a quadratic form :

Lemma 2.1 *Let $\varphi \in C^\infty(X;\mathbb{R})$ and let $x_0 \in X$ be a non-degenerate critical point. Then there are neighborhoods U of $0 \in \mathbb{R}^n$ and V of x_0 and a C^∞ diffeomorphism $\mathcal{H} : V \to U$ such that*

$$\varphi \circ \mathcal{H}^{-1}(x) = \varphi(x_0) + \frac{1}{2}(x_1^2 + \ldots + x_r^2 - x_{r+1}^2 - \ldots - x_n^2).$$

Here $(r, n-r)$ is the signature of $\varphi''(x_0)$ (so that r, $n-r$ are respectively the number of positive and negative eigenvalues).

Proof: After a translation and a linear change of coordinates, we may assume that $x_0 = 0$ and that

$$\varphi(x) = \frac{1}{2}(x_1^2 + \ldots + x_r^2 - x_{r+1}^2 - \ldots - x_n^2) + \mathcal{O}(|x|^3), \quad x \to 0.$$

By Taylor's formula,

$$\begin{aligned}\varphi(x) &= \int_0^1 (1-t)\frac{\partial^2}{\partial t^2}(\varphi(tx))\,dt = \frac{1}{2}\sum\sum q_{j,k}(x)\,x_j\,x_k \\ &= \frac{1}{2}\langle x, Q(x)\,x\rangle,\end{aligned}$$

where $Q(x) = (q_{j,k}(x))$, $q_{j,k}(x) = 2\int_0^1 (1-t)\frac{\partial^2 \varphi}{\partial x_j \partial x_k}(tx)\,dt$, $q_{j,k}(0) = \frac{\partial^2 \varphi}{\partial x_j \partial x_k}(0)$,

$$Q(0) = \begin{pmatrix} 1 & & & & & 0 \\ & \ddots & & & & \\ & & 1 & & & \\ & & & -1 & & \\ & & & & \ddots & \\ 0 & & & & & -1 \end{pmatrix}.$$

We look for $\mathcal{H}$ of the form $\mathcal{H}(x) = A(x)\,x$, where the matrix $A(x)$ depends smoothly on x and satisfies $A(0) = I$. Then $A(x)$ should satisfy $\langle x, Q(x)\,x\rangle = \langle A(x)\,x, Q(0)\,A(x)\,x\rangle$. It then suffices to have $Q(x) = {}^t\!A(x)\,Q(0)\,A(x)$.

Let $\operatorname{Sym}(n, \mathbb{R})$ be the space of real symmetric $n \times n$ matrices and consider the map

$$\mathcal{F} : \operatorname{Mat}(n, \mathbb{R}) \ni A \mapsto {}^t\!A\,Q(0)\,A \in \operatorname{Sym}(n, \mathbb{R}).$$

where $\operatorname{Mat}(n, \mathbb{R})$ denotes the space of all real $n \times n$ matrices. The differential at the point $A = I$ is

$$d\mathcal{F} : \operatorname{Mat}(n, \mathbb{R}) \ni \delta A \mapsto {}^t(\delta A)\,Q(0) + Q(0)\,(\delta A) \in \operatorname{Sym}(n, \mathbb{R}).$$

$d\mathcal{F}$ is surjective, for if $\delta B \in \operatorname{Sym}(n, \mathbb{R})$, then $\delta A = \frac{1}{2} Q(0)^{-1}\,\delta B$ is a solution to $d\mathcal{F}(\delta A) = \delta B$.

By the implicit function theorem, $\mathcal{F}$ has a local smooth right inverse $\mathcal{G}$ mapping a neighborhood of zero in $\operatorname{Sym}(n, \mathbb{R})$ into a neighborhood of 0 in $\operatorname{Mat}(n, \mathbb{R}) : \mathcal{F}\mathcal{G} = \operatorname{id}$. We then get A with the required properties by taking $A(x) = \mathcal{G}(Q(x))$. The map $\mathcal{H}(x) = A(x)\,x$ is then a diffeomorphism from a neighborhood of 0 onto a neighborhood of 0, since $d\mathcal{H}(0) = A(0) = I$. $\square$

If $u \in L^1(\mathbb{R}^n)$ we define its Fourier transform $\mathcal{F}u(\xi) = \hat{u}(\xi) = \int e^{-ix\xi} u(x)\,dx$ and we extend this definition to the space $\mathcal{S}'(\mathbb{R}^n)$ of temperate distributions the usual way. In one variable, the Fourier transform of $e^{-x^2/2}$ is $(2\pi)^{\frac{1}{2}} e^{-\xi^2/2}$, and by scaling and analytic continuation we get for $z \in \dot{\mathbb{C}}$, $\operatorname{Re} z \geq 0$:

$$\mathcal{F}(e^{-zx^2/2})(\xi) = \left(\frac{2\pi}{z}\right)^{\frac{1}{2}} e^{-\xi^2/2z} \quad (\text{in } \mathcal{S}'(\mathbb{R}^n_\xi)),$$

with the natural branch of the square root in the right half-plane. In particular

(2.3) $$\mathcal{F}(e^{\pm ix^2/2}) = (2\pi)^{\frac{1}{2}} e^{\pm i\pi/4} e^{\mp i\xi^2/2}.$$

If $Q \in \operatorname{Sym}(n, \mathbb{R})$ is non-degenerate with r positive and $n - r$ negative eigenvalues we put $\operatorname{sgn} Q = r - (n - r)$. Combining (2.3) and a linear change of variables, we get

(2.4) $$\mathcal{F}: e^{i\langle x, Qx\rangle/2} \mapsto (2\pi)^{\frac{n}{2}} e^{i\frac{\pi}{4}\operatorname{sgn} Q} |\det Q|^{-\frac{1}{2}} e^{-i\langle \xi, Q^{-1}\xi\rangle/2}.$$

If $u \in C_0^\infty(\mathbb{R}^n)$, Parseval's formula gives

(2.5) $$\int e^{i\lambda\langle x, Qx\rangle/2} u(x)\, dx =$$
$$(2\pi)^{-\frac{n}{2}} \lambda^{-\frac{n}{2}} |\det Q|^{-\frac{1}{2}} e^{i\frac{\pi}{4}\operatorname{sgn} Q} \int e^{-i\langle \xi, Q^{-1}\xi\rangle/2\lambda} \hat{u}(\xi)\, d\xi.$$

By Taylor's formula $\left| e^{it} - \sum_{0}^{N-1} \frac{(it)^k}{k!} \right| \leq \frac{|t|^N}{N!}$ for real t, and hence,

$$e^{-i\langle\xi, Q^{-1}\xi\rangle/2\lambda} = \sum_{0}^{N-1} \frac{1}{k!}\left(\frac{1}{2\lambda i}\langle\xi, Q^{-1}\xi\rangle\right)^k + R_N(\xi, \lambda),$$

where
$$|R_N(\xi, \lambda)| \leq (2\lambda)^{-N} |\langle\xi, Q^{-1}\xi\rangle|^N / N!.$$

Also,
$$\int \langle\xi, Q^{-1}\xi\rangle^k \hat{u}(\xi)\, d\xi = (2\pi)^n \left(\langle D_x, Q^{-1}D_x\rangle^k u\right)(0).$$

Hence for $u \in C_0^\infty(\mathbb{R}^n)$:

(2.6) $$\int e^{i\lambda\langle x, Qx\rangle/2} u(x)\, dx =$$
$$\sum_{k=0}^{N-1} \frac{(2\pi)^{\frac{n}{2}} e^{i\frac{\pi}{4}\operatorname{sgn} Q}}{k! |\det Q|^{\frac{1}{2}} \lambda^{k+\frac{n}{2}}} \left(\frac{1}{2i}\langle D_x, Q^{-1}D_x\rangle\right)^k u(0) + S_N(u, \lambda),$$

where

(2.7) $$|S_N(u, \lambda)| \leq \frac{C_Q}{N!}(2\pi)^{-\frac{n}{2}} \lambda^{-\frac{n}{2}-N} \int \left|\left(\frac{1}{2}\langle\xi, Q^{-1}\xi\rangle\right)^N \hat{u}(\xi)\right| d\xi$$
$$\leq \tilde{C}_Q (N!)^{-1} \lambda^{-N-\frac{n}{2}} \sum_{|\alpha|\leq n+1} \left\| D^\alpha \left(\frac{1}{2}\langle D, Q^{-1}D\rangle\right)^N u \right\|_{L^1(\mathbb{R}^n)}.$$

Here the last expression may be replaced by

$$C_{Q,\varepsilon}(N!)^{-1}\lambda^{-N-\frac{n}{2}}\left\|\left(\tfrac{1}{2}\langle D,Q^{-1}D\rangle\right)^N u\right\|_{H^{\frac{n}{2}+\varepsilon}(\mathbb{R}^n)},$$

for any $\varepsilon > 0$.

Example 2.2 We replace n by $2n$ and x by (x,y), $x,y \in \mathbb{R}^n$ and we put $Q = Q^{-1} = \begin{pmatrix} 0 & -I \\ -I & 0 \end{pmatrix}$. Then

$$\tfrac{1}{2}\langle (x,y), Q(x,y)\rangle = -x\cdot y, \quad \tfrac{1}{2i}\langle D_{(x,y)}, Q^{-1}D_{(x,y)}\rangle = \tfrac{1}{i}\sum \partial_{x_j}\partial_{y_j}.$$

Then (2.6), (2.7) become

(2.6)′
$$\left(\tfrac{\lambda}{2\pi}\right)^n \iint e^{-i\lambda x\cdot y} u(x,y)\,dx\,dy$$
$$= \sum_{k=0}^{N-1} \tfrac{1}{k!\lambda^k}\left(\left(\tfrac{1}{i}\sum_1^n \partial_{x_j}\partial_{y_j}\right)^k u\right)(0,0) + S_N(u,\lambda)$$
$$= \sum_{|\alpha|\le N-1} \tfrac{1}{\lambda^{|\alpha|}\alpha! i^{|\alpha|}} (\partial_x^\alpha \partial_y^\alpha u)(0,0) + S_N(u,\lambda)$$

where

(2.7)′ $$|S_N(u,\lambda)| \le \tfrac{C}{N!\lambda^N}\sum_{|\alpha+\beta|\le 2n+1}\|\partial_x^\alpha \partial_y^\beta (\partial_x\cdot \partial_y)^N u\|_{L^1}.$$

This example is the basic ingredient of the classical calculus of pseudodifferential operators.

Combining the Morse lemma with (2.6), (2.7), we get

Proposition 2.3 Let $\varphi \in C^\infty(X;\mathbb{R})$ have the non-degenerate critical point $x_0 \in X$ and assume that $\varphi'(x) \ne 0$ for $x \ne x_0$. Then there are differential operators $A_{2\nu}(D)$ of order $\le 2\nu$ such that for every compact $K \subset X$ and every $N \in \mathbb{N}$, there is a constant $C = C_{K,N}$ such that for every $u \in C^\infty(X) \cap \mathcal{E}'(K)$

$$\left|\int e^{i\lambda\varphi(x)}u(x)\,dx - \left(\sum_0^{N-1}(A_{2\nu}(D_x)u)(x_0)\lambda^{-\nu-\frac{n}{2}}\right)e^{i\lambda\varphi(x_0)}\right|$$
$$\le C\lambda^{-N-\frac{n}{2}}\sum_{|\alpha|\le 2N+n+1}\sup|\partial^\alpha u(x)|, \quad \lambda \ge 1.$$

Moreover $A_0 = \dfrac{(2\pi)^{\frac{n}{2}} \cdot e^{i\frac{\pi}{4}\operatorname{sgn}\varphi''(x_0)}}{|\det \varphi''(x_0)|^{\frac{1}{2}}}.$

Recall that a differential operator of order $\leq m$ on an open set $X \subset \mathbb{R}^n$ is an operator of the form $P(x, D_x) = \sum_{\alpha \in \mathbb{N}^n, |\alpha| \leq m} a_\alpha(x) D_x^\alpha$, $(|\alpha| = \|\alpha\|_{\ell^1})$.

Proof: Let V be a neighborhood of x_0 as in Lemma 2.1, and let $\chi \in C_0^\infty(V)$ be equal to 1 near x_0. Then $\int e^{i\lambda\varphi} \chi u \, dx$ can be reduced to an integral of the form (2.5) (with a new u) by means of the Morse lemma. The value of the Jacobian of the change of variables at 0 can be determined from the relation $\varphi''(x_0) = {}^t d\mathcal{H}(0) \circ Q(0) \circ d\mathcal{H}(0)$ (in the notations of Lemma 2.1 and its proof). The integral $\int e^{i\lambda\varphi(x)} (1 - \chi(x)) u(x) \, dx$ can be treated by means of repeated integrations by parts. □

In practice φ, u will often depend smoothly on some additional parameters $t \in T \subset \mathbb{R}^k$. Then (under reasonable assumptions), $x_0 = x_0(t)$, $A_{2\nu}(D_x) = A_{2\nu}(t, D_x)$ will depend smoothly on t, and $C_{K,N}$ will be independent of t for t in any fixed compact set $L \subset T$. We leave the precise formulation of such a result to the reader.

There are many variants and extensions of the method of stationary phase. Instead of isolated critical points we may treat the case when the critical points form a submanifold (under suitable assumptions of non-degeneracy). One may also treat the case when φ is complex-valued and in particular when φ is analytic one can establish the link with the method of steepest descent. One may use infinitesimal versions of the Morse lemma when deforming φ and replace the change of variables in the Morse lemma by integration by parts by means of a suitable vector field.

The following variant of the discussion leading to (2.6), (2.7) uses more explicitly a certain Schrödinger operator. We consider for $t \searrow +0$:

$$I(t, u) = \frac{1}{t^{\frac{n}{2}}} \int e^{i\langle x, Qx\rangle/2t} u(x) \, dx,$$

$u \in C_0^\infty(\mathbb{R}^n)$, $Q \in \mathrm{Sym}(n, \mathbb{R})$, $\det Q \neq 0$. If $P = \frac{1}{2i} \langle D_x, Q^{-1} D_x \rangle$ we notice that

$$\frac{\partial}{\partial t} \left(\frac{1}{t^{\frac{n}{2}}} e^{i\langle x, Qx\rangle/2t} \right) = P(D_x) \left(\frac{1}{t^{\frac{n}{2}}} e^{i\langle x, Qx\rangle/2t} \right).$$

Hence,

(2.8) $\qquad \dfrac{d}{dt} I(t, u) = I(t, Pu), \qquad \dfrac{d^N}{dt^N} I(t, u) = I(t, P^N u).$

Moreover, for $u \in C_0^\infty$ we find, for instance by using (2.5), that

(2.9) $\qquad I(t, u) \longrightarrow \dfrac{(2\pi)^{\frac{n}{2}} e^{i\frac{\pi}{4} \mathrm{sgn} Q}}{|\det Q|^{\frac{1}{2}}} u(0), \quad t \to 0,$

(2.10) $$|I(t,u)| \leq C_n \frac{(2\pi)^{\frac{n}{2}}}{|\det Q|^{\frac{1}{2}}} \sum_{|\alpha| \leq n+1} \|\partial^\alpha u\|_{L^1}.$$

Combining (2.8)–(2.10) with a Taylor expansion at $t = 0$, we get

$$I(t,u) = \frac{(2\pi)^{\frac{n}{2}} e^{i\frac{\pi}{4}\operatorname{sgn} Q}}{|\det Q|^{\frac{1}{2}}} \Big(\sum_{0}^{N-1} \frac{t^k}{k!} (P^k u)(0) + R_N(u,t)\Big),$$

where

$$|R_N(u,t)| \leq \frac{t^N}{N!} C_{n,Q} \sum_{|\alpha| \leq n+1} \|\partial^\alpha P^N u\|_{L^1}.$$

Exercises

Exercise 2.1 (Another proof of the Morse lemma)
Let $\varphi \in C^\infty(\Omega, \mathbb{R})$ with $x_0 \in \Omega$ a non-degenerate critical point : $\varphi'(x_0) = 0$, $\det \varphi''(0) \neq 0$. Suppose for simplicity that $x_0 = 0$ and $\varphi(x_0) = 0$.

a) Show that for x in some neighborhood of 0, $\varphi(x) = \frac{1}{2} \sum_{1 \leq i,j \leq n} q_{ij}(x) x_i x_j$, with $q_{ij} = q_{ji} \in C^\infty$ and $(q_{ij}(0))$ non-degenerate.

b) Show that after a linear change of variables, it is possible to suppose $q_{11}(0) \neq 0$. Then show that

$$\varphi(x) = \frac{1}{2}\Big(\varepsilon_1(\ell_1(x))^2 + \sum_{2 \leq i,j \leq n} \tilde{q}_{ij}(x) x_i x_j\Big)$$

with $\varepsilon_1 = \pm 1$, $\ell_1(x) = \sum_{1 \leq j \leq n} a_{1,j}(x) x_j$ and $a_{1,j}, \tilde{q}_{ij} \in C^\infty$.

c) Finally show that

$$\varphi(x) = \frac{1}{2} \sum_{i=1}^n \varepsilon_i \big(\ell_i(x)\big)^2$$

with $\ell_i(x) = \sum a_{ij}(x) x_j$, $\varepsilon_i = \pm 1$, $a_{ij} \in C^\infty$.

d) Show that $(x_1, \ldots, x_n) \mapsto (\ell_1(x), \ldots, \ell_n(x))$ is a diffeomorphism $U \to V$, U and V neighborhoods of 0.

e) Derive the Morse lemma.

Exercise 2.2
a) Do the following functions converge in $\mathcal{D}'(\mathbb{R})$ as $\lambda \to +\infty$?
$$u_\lambda(x) = \lambda^N e^{-i\lambda x}$$
$$v_\lambda(x) = \lambda^{\frac{1}{2}} e^{-i\lambda \frac{x^2}{2}}$$
$$w_\lambda(x) = \lambda^{\frac{1}{2}} e^{+i\lambda \frac{x^2}{2}}$$

b) Let $f \in C^\infty(\mathbb{R}, \mathbb{R})$ with $f'(x) \neq 0$, $\forall x \in \mathbb{R}$. Then same question for
$$u_\lambda(x) = \lambda^N e^{-i\lambda f(x)}$$
$$v_\lambda(x) = \lambda^{\frac{1}{2}} e^{-i\lambda \frac{(f(x))^2}{2}}$$

Exercise 2.3
Let $\chi \in \mathcal{S}(\mathbb{R})$, $\chi(0) = 1$.
a) Show the existence of a limit as $\varepsilon \to 0$ of
$$\int e^{-i\lambda(y+\frac{y^3}{3})} \chi(\varepsilon y)\, dy, \quad \lambda \in \mathbb{R}\setminus\{0\}.$$
Show that the limit $I(\lambda)$ is a C^∞ function of λ.
Hint : choose a suitable differential operator L such that $L\bigl(e^{-i\lambda(y+\frac{y^3}{3})}\bigr) = e^{-i\lambda(y+\frac{y^3}{3})}$.

b) Show that for every N, $|I(\lambda)| \leq C_N |\lambda|^{-N}$ if $|\lambda| > 1$.

c) Show that $J(\lambda) = \lim_{\varepsilon \to 0} \int e^{-i\lambda(y-\frac{y^3}{3})} \chi(\varepsilon y)\, dy$ does exist for $\lambda \in \mathbb{R}\setminus\{0\}$.

d) Find the asymptotics of $J(\lambda)$ as $\lambda \to \pm\infty$.

Exercise 2.4
The following problem appears in the method of steepest descent. Show that for $u \in C_0^\infty(\mathbb{R}^n)$, $\lambda \geq 1$:
$$\int e^{-\lambda x^2/2} u(x)\, dx = \sum_{k=0}^{N-1} \frac{(2\pi)^{\frac{n}{2}}}{k!\, \lambda^{k+\frac{n}{2}}} \left(\bigl(\frac{1}{2}\Delta\bigr)^k u\right)(0) + S_N(u, \lambda),$$
where $|S_N(u, \lambda)| \leq C_{n,N}\, \lambda^{-N-\frac{n}{2}} \sum_{|\alpha|=2N} \sup |\partial^\alpha u(x)|$.

Exercise 2.5 (Stirling's formula)
Let $F(\lambda) = \Gamma(\lambda + 1) = \int_0^{+\infty} e^{-t} t^\lambda\, dt$, $\lambda \geq 1$ ($F(n) = n!$, $n \in \mathbb{N}$).
One wants to find the asymptotics of $F(\lambda)$ as $\lambda \to +\infty$.

a) Rewrite the integral by means of the change of variable $t = \lambda(1+s)$.

b) Use Exercise 2.4 and show

$$F(\lambda) = \left(\frac{\lambda}{e}\right)^\lambda \sqrt{2\pi\lambda}\,(1 + a_1\lambda^{-1} + a_2\lambda^{-2} + \ldots).$$

For $\lambda = n \in \mathbb{N}$ deduce Stirling's formula.

c) Calculate a_1 and a_2.

Notes

For the Morse lemma and the method of stationary phase, we have followed the presentation in Hörmander [Hö2].

In the analytic category the variant, called the method of steepest descent, is a powerful tool which also is of use in the C^∞ setting. See Melin–Sjöstrand [MS] and Sjöstrand [S] in the multidimensional case.

3 Pseudodifferential operators

Let $0 < \rho \leq 1$, $0 \leq \delta < 1$, X open $\subset \mathbb{R}^n$. A pseudodifferential operator is a Fourier integral operator $A : C_0^\infty(X) \to \mathcal{D}'(X)$ of the form

$$(3.1) \quad Au(x) = \frac{1}{(2\pi)^n} \iint e^{i(x-y)\theta} a(x,y,\theta) u(y) \, dy \, d\theta, \quad u \in C_0^\infty(X),$$

where $a \in S_{\rho,\delta}^m(X \times X \times \mathbb{R}^n)$. We denote by $L_{\rho,\delta}^m(X)$ the space of these operators and we say that $A \in L_{\rho,\delta}^m(X)$ is of order $\leq m$ and of type (ρ, δ).

Example 3.1 If A is a differential operator on X of order $\leq m$ with smooth coefficients, then $A \in L_{1,0}^m(X)$. In fact, if $A = \sum_{|\alpha| \leq m} a_\alpha(x) D_x^\alpha$, $a_\alpha \in C^\infty(X)$, then by Fourier's inversion formula we get

$$Au(x) = \int e^{ix\xi} a(x,\xi) \hat{u}(\xi) \frac{d\xi}{(2\pi)^n} = \iint e^{i(x-y)\xi} a(x,\xi) u(y) \, dy \, \frac{d\xi}{(2\pi)^n},$$

$a(x,\xi) = \sum_{|\alpha| \leq m} a_\alpha(x) \xi^\alpha \in S_{1,0}^m(X \times \mathbb{R}^n)$. Here the last integral first makes sense as an iterated one, but inserting $\chi(\varepsilon\xi)$, $\chi \in C_0^\infty$, $\chi(0) = 1$ and letting ε tend to 0, we see that the distribution kernel of A is given by the oscillatory integral

$$K_A(x,y) = \int e^{i(x-y)\xi} a(x,\xi) \frac{d\xi}{(2\pi)^n}.$$

Example 3.2 The inverse of $(1 - \Delta) : \mathcal{S}(\mathbb{R}^n) \to \mathcal{S}(\mathbb{R}^n)$, where $\mathcal{S}(\mathbb{R}^n)$ is the Schwartz space of test functions and $-\Delta = -\sum \frac{\partial^2}{\partial x_j^2} = \sum D_{x_j}^2$, is given by

$$Au(x) = (1-\Delta)^{-1} u(x) = \int e^{ix\xi} \frac{1}{(1+\xi^2)} \hat{u}(\xi) \frac{d\xi}{(2\pi)^n} \quad (\xi^2 = \sum \xi_j^2).$$

Hence $A \in L_{1,0}^{-2}(\mathbb{R}^n)$, and it is not a differential operator. From our generalities on FIO's we get

1) If $K_A \in \mathcal{S}(X \times X)$ is the distribution kernel of $A \in L_{\rho,\delta}^m(X)$, then $\operatorname{sing\,supp}(K_A) \subset \Delta(X \times X) \stackrel{\text{def}}{=} \{(x,x) \in X \times X\}$ (= "the diagonal of $X \times X$"). In fact $(x-y)\theta$ is a phase function and $d_\theta((x-y)\theta)$ vanishes precisely when $x = y$.

2) Since $d_{y,\theta}((x-y)\theta) \neq 0$ and $d_{x,\theta}((x-y)\theta) \neq 0$ when $\theta \neq 0$, the pseudodifferential operators in $L_{\rho,\delta}^m(X)$ are continuous $C_0^\infty(X) \to C^\infty(X)$ and have (unique) continuous extensions $\mathcal{E}'(X) \to \mathcal{D}'(X)$. (We will not distinguish between these extensions and the original operators.)

From 1) and 2) we deduce when $A \in L^m_{\rho,\delta}(X)$:

3) sing supp $Au \subset$ sing supp u, $\forall u \in \mathcal{E}'(X)$. In fact, let $u \in \mathcal{E}'(X)$, $x_0 \in X\backslash$sing supp(u). Then we can find $\varphi, \psi \in C_0^\infty(X)$ such that $\varphi = 1$ near x_0, $\psi = 1$ near sing supp(u). (By "near" we mean "in some neighborhood of"). Then $Au \equiv A\psi u$ mod $C^\infty(X)$ according to 2) and $\varphi A\psi u \equiv 0$ mod $C^\infty(X)$ according to 1) since the distribution kernel of $\varphi A\psi$ is $\varphi(x)\psi(y) K_A(x,y) \in C^\infty(X \times X)$.

Here we pause and recall a general fact of distribution theory. If $X \subset \mathbb{R}^{n_X}$, $Y \subset \mathbb{R}^{n_Y}$ are open and $A : C_0^\infty(Y) \to \mathcal{D}'(X)$ is continuous and linear with distribution kernel $K_A(x,y) \in \mathcal{D}'(X \times Y)$, then the following two statements are equivalent :

(i) A (extends to an operator which) is continuous $\mathcal{E}'(Y) \to C^\infty(X)$.

(ii) $K_A \in C^\infty(X \times Y)$.

If A satisfies (i) or (ii) we say that A is smoothing.

4) If $X = Y$ we denote by $L^{-\infty}(X)$ the space of operators of the form (3.1) with $a \in S^{-\infty}(X \times X \times \mathbb{R}^n)$. If $A \in L^{-\infty}(X)$ we see that $K_A \in C^\infty(X \times X)$ so A is smoothing. Conversely if A is smoothing, then A is of the form (3.1) with $a(x,y,\theta) = K_A(x,y) e^{-i(x-y)\theta} \chi(\theta) \in S^{-\infty}$, if $\chi \in C_0^\infty(\mathbb{R}^n)$, $\int \chi(\theta) d\theta = (2\pi)^n$. Hence $L^{-\infty}(X)$ is the space of smoothing operators $\mathcal{E}'(X) \to C^\infty(X)$.

In order to simplify the discussion of composition of pseudodifferential operators and Fourier integral operators it is sometimes convenient to introduce the notion of properly supported operators. Let $X \subset \mathbb{R}^{n_X}$, $Y \subset \mathbb{R}^{n_Y}$ be open sets. If C is a closed subset of $X \times Y$, we say that C is proper if the two projections

$$\Pi_x : C \ni (x,y) \mapsto x \in X$$
$$\Pi_y : C \ni (x,y) \mapsto y \in Y$$

are proper (i.e. the inverse image of every compact subset of X and Y respectively is compact). If we view C as the graph of a relation $Y \to X$ (which we also denote by C), then C is proper if and only if

(3.2) $\qquad C(K) = \{x \in X \,;\, \exists y \in K \text{ s.t. } (x,y) \in C\}$

and

(3.3) $\qquad C^{-1}(L) = \{y \in Y \,;\, \exists x \in L \text{ s.t. } (x,y) \in C\}$

are compact for all compact sets $K \subset X$ and $L \subset Y$.

An operator $A : C_0^\infty(Y) \to \mathcal{D}'(X)$ is said to be properly supported if $\operatorname{supp} K_A \subset X \times Y$ is proper. Viewing $C = \operatorname{supp} K_A$ as a relation we see that $\operatorname{supp} Au \subset C(\operatorname{supp} u)$ when $u \in C_0^\infty(Y)$, so if A is properly supported, then A is continuous $C_0^\infty(Y) \to \mathcal{E}'(X)$. Using also the property (3.3), we see that if A is properly supported, then A has a (unique) continuous extension $\tilde{A} : C^\infty(Y) \to \mathcal{D}'(X)$. In fact, the uniqueness is obvious as usual and for the existence it suffices to put $\tilde{A}u = A\chi_{\tilde{X}} u$ on any given $\tilde{X} \subset\subset X$ with $\chi_{\tilde{X}} \in C_0^\infty(Y)$ equal to 1 near $C^{-1}(\overline{\tilde{X}})$. (This does not depend on the choice of $\chi_{\tilde{X}}$.) We shall not distinguish between A and $\tilde{A}$.

If $A \in L_{\rho,\delta}^m(X)$ is properly supported, then A is continuous :

$$C_0^\infty(X) \to C_0^\infty(X), \ C^\infty(X) \to C^\infty(X)$$
$$\mathcal{E}'(X) \to \mathcal{E}'(X), \ \mathcal{D}'(X) \to \mathcal{D}'(X).$$

If $B \in L_{\rho,\delta}^{m'}(X)$ is a second operators not necessarily properly supported, then $A \circ B$ and $B \circ A$ are well-defined operators $C_0^\infty(X) \to C^\infty(X)$, $\mathcal{E}(X) \to \mathcal{D}'(X)$. More generally we can compose finitely many pseudodifferential operators on X provided that all, except possibly one, are properly supported. If all the factors are properly supported then so is the product.

We shall next construct $\chi \in C^\infty(X \times X)$ such that $\chi = 1$ in a neighborhood of $\Delta(X \times X)$ and such that $\operatorname{supp} \chi$ is proper. Let $1 = \sum_0^\infty \varphi_j(x)$, $\varphi_j(x) \in C_0^\infty(X)$ be a locally finite partition of unity (if $K \subset X$ is a compact set then there are only finitely many of the φ_j with $K \cap \operatorname{supp} \varphi_j \neq \emptyset$). Then $1 = \sum \sum \varphi_j(x) \varphi_k(y)$ is a locally finite partition of unity on $X \times X$. We put

$$\chi(x,y) = \sum\sum_{\operatorname{supp}\varphi_j \cap \operatorname{supp}\varphi_k \neq \emptyset} \varphi_j(x)\varphi_k(y).$$

Then

$$1 - \chi(x,y) = \sum\sum_{\operatorname{supp}\varphi_j \cap \operatorname{supp}\varphi_k = \emptyset} \varphi_j(x)\varphi_k(y)$$

vanishes in a neighborhood of the diagonal.

On the other hand, if $C = \operatorname{supp}\chi$ is viewed as a relation and if $K \subset X$ is compact, then

$$C(K) \subset \bigcup \operatorname{supp}\varphi_j,$$

where the union is taken over all j such that there exists $k = k(j)$, with $\operatorname{supp}\varphi_j \cap \operatorname{supp}\varphi_k \neq \emptyset$, $\operatorname{supp}\varphi_k \cap K \neq \emptyset$. It is therefore clear that $C(K)$ is compact and similarly $C^{-1}(K)$ is compact.

Remark 3.3 Every $A \in L_{\rho,\delta}^m(X)$ has a decomposition $A = A' + A''$, where $A' \in L_{\rho,\delta}^m(X)$ is properly supported and $A'' \in L^{-\infty}$. In fact, we define A', A'' by introducing the cut-offs $\chi(x,y)$ and $1 - \chi(x,y)$ into the integral (3.1).

The next result permits us to eliminate the dependence on y in the symbol $a(x, y, \theta)$ in (3.1).

Theorem 3.4 *Let $A \in L_{\rho,\delta}^m(X)$ be properly supported of the form (3.1). We also assume that $\rho > \delta$. Then $b(x,\xi) \stackrel{\text{def}}{=} e^{-ix\xi} A(e^{i(\cdot)\xi})$ belongs to $S_{\rho,\delta}^m(X \times \mathbb{R}^n)$ and has the asymptotic development*

$$b(x,\xi) \sim \sum_{\alpha \in \mathbb{N}^n} \frac{i^{-|\alpha|}}{\alpha!} (\partial_\xi^\alpha \partial_y^\alpha a(x,y,\xi))|_{y=x}.$$

Moreover, $Au(x) = \frac{1}{(2\pi)^n} \int e^{ix\xi} b(x,\xi) \hat{u}(\xi) d\xi$, $u \in C_0^\infty(X)$ and we call $b(x,\xi)$ the (complete) symbol of A. We write $b = \sigma_A$. (The asymptotic sum above is defined by regrouping terms with the same value of $|\alpha|$ and we notice that $\partial_\xi^\alpha \partial_y^\alpha a \in S_{\rho,\delta}^{m-|\alpha|(\rho-\delta)}$.)

Proof: The theorem is trivially valid when $A \in L^{-\infty}$ and hence after introducing a cut-off $\chi(x,y)$ with the properties above, we may assume that $\Pi_{(x,y)} \text{supp}\, a = \{(x,y) \in X \times X; \exists \theta \in \mathbb{R}^n, (x,y,\theta) \in \text{supp}\, a\}$ is proper. Then,

$$e^{-ix\xi} A(e^{i(\cdot)\xi}) = \iint a(x,y,\theta) e^{i(x-y)(\theta-\xi)} \frac{dy\, d\theta}{(2\pi)^n} \stackrel{\text{def}}{=} b(x,\xi).$$

The integral is the limit when approximating a by a sequence of symbols in $S^{-\infty}$ but can also be interpreted as an iterated integral. The only critical point of the phase $(y,\theta) \mapsto (x-y)(\theta-\xi)$ is given by $y = x$, $\theta = \xi$.

Let $\chi \in C_0^\infty([0, +\infty[; [0,1])$ be equal to 1 on $\left[0, \frac{1}{3}\right]$ and have its support in $\left[0, \frac{1}{2}\right[$. For $|\xi| \geq 2$ and for x in some fixed compact set $K \subset X$, we put

$$b_2(x,\xi) = \iint a(x,y,\theta) \left(1 - \chi\left(\frac{|\theta-\xi|}{|\xi|}\right)\right) e^{i(x-y)(\theta-\xi)} \frac{dy\, d\theta}{(2\pi)^n}.$$

In this integral we integrate by parts, using

$$L = \frac{1}{|\theta-\xi|^2} \sum_1^n (\xi_j - \theta_j) D_{y_j}.$$

In the support of $1 - \chi\left(\frac{|\theta-\xi|}{|\xi|}\right)$ we have $|\theta - \xi| \sim 1 + |\theta| + |\xi|$ (where we also use "~" for the relation "same order of magnitude", $a \sim b$ if there is a constant $C > 0$ (independent of the parameters) such that $\frac{a}{C} \leq b \leq Ca$).

We now use that ${}^tL = -L$ and (with $\mathcal{O}_k(1)$ indicating a quantity uniformly bounded in θ, ξ):

$$L^k(a(1-\chi)) = \mathcal{O}_k(1) \frac{(1+|\theta|)^{\delta k + m}}{(1+|\theta|+|\xi|)^k} = \mathcal{O}_k(1) (1+|\theta|+|\xi|)^{m-(1-\delta)k}$$
$$= \mathcal{O}_k(1)(1+|\theta|)^{-(n+1)}(1+|\xi|)^{m+n+1-(1-\delta)k},$$

if $k \in \mathbb{N}$ and $m + n + 1 - (1 - \delta) k \leq 0$. Hence $b_2(x, \xi) = \mathcal{O}_k(1) (1 + |\xi|)^{m+n+1-(1-\delta)k}$ for every $k > 0$, assuming for simplicity that $m \geq 0$. We have similar estimates for the derivatives of b_2, and deduce that $b_2 \in S^{-\infty}$.

It remains to study for $|\xi| \geq 1$, $x \in K$ compact $\subset X$:

$$b_1(x, \xi) = \iint a(x, y, \theta) \chi\left(\frac{|\theta - \xi|}{|\xi|}\right) e^{i(x-y)(\theta-\xi)} \frac{dy\, d\theta}{(2\pi)^n}$$
$$= \iint a(x, x+s, \xi+\sigma) \chi\left(\frac{|\sigma|}{|\xi|}\right) e^{-is\sigma} \frac{ds\, d\sigma}{(2\pi)^n}.$$

In order to apply the method of stationary phase, we put $\xi = \lambda\omega$, $\lambda = |\xi|$ and make the change of variables $\sigma = \lambda\tilde\sigma$ and obtain after dropping the $\tilde{}$:

$$b_1(x, \lambda\omega) = \left(\frac{\lambda}{2\pi}\right)^n \iint a(x, x+s, \lambda(\omega+\sigma)) \chi(|\sigma|) e^{-i\lambda s\sigma} ds\, d\sigma.$$

We can then apply stationary phase as in Example 2.2 and obtain

$$b_1(x, \lambda\omega) = \sum_{|\alpha| \leq N-1} \frac{\lambda^{-|\alpha|} i^{-|\alpha|}}{\alpha!} \partial_s^\alpha \partial_\sigma^\alpha (a(x, x+s, \lambda(\omega+\sigma))\chi(|\sigma|))|_{s=\sigma=0} + S_N(\lambda)$$
$$= \sum_{|\alpha| \leq N-1} \frac{i^{-|\alpha|}}{\alpha!} (\partial_y^\alpha \partial_\xi^\alpha a(x, y, \xi))|_{y=x} + S_N(\lambda),$$

where

$$|S_N(\lambda)| \leq \frac{C_{N,K}}{\lambda^N} \sum_{|\alpha+\beta| \leq 2n+1} \sup_{s,\sigma} \left|\partial_s^\alpha \partial_\sigma^\beta (\partial_s \cdot \partial_\sigma)^N (a(x, x+s, \lambda(\omega+\sigma)) \chi(|\sigma|))\right|.$$

Here $(\partial_s \cdot \partial_\sigma)^N (a(x, x+s, \lambda(\omega+\sigma)) \chi(|\sigma|))$ is a finite sum of terms of the form $\lambda^{|\beta'|}(\partial_y^{\alpha'} \partial_\xi^{\beta'} a)(x, x+s, \lambda(\omega+\sigma)) \partial_\sigma^{\beta''} \chi(|\sigma|)$, with $\beta' + \beta'' = \alpha'$, $|\alpha'| = N$. The derivative $\partial_s^\alpha \partial_\sigma^\beta$ with $|\alpha + \beta| \leq 2n+1$ of such a term is bounded in absolute value by

$$C \cdot \lambda^{(2n+1)\max(\delta, 1-\rho) + m + |\beta'|(1-\rho) + |\alpha'|\delta} \leq C\lambda^{m+(2n+1)\max(\delta, 1-\rho) + N - (\rho-\delta)N}.$$

Hence $|S_N(\lambda)| \leq C_N |\xi|^{m+(2n+1)\max(\delta, 1-\rho) - (\rho-\delta)N}$, $x \in K$ compact $\subset X$, $|\xi| \geq 1$.

We also check that every derivative of $b_1(x, \xi)$ is of temperate growth when $|\xi| \to \infty$ (uniformly for x in any fixed compact subset of X), so that Proposition 1.9 applies and we then deduce that $b_1(x, \xi) \in S_{\rho,\delta}^m$ and has the desired asymptotic expansion. Since $b(x, \xi) \equiv b_1(x, \xi)$ mod $S^{-\infty}$, we obtain the desired properties for b.

If $u \in C_0^\infty(X)$ we write $u(x) = \int e^{ix\xi} \hat{u}(\xi) \frac{d\xi}{(2\pi)^n}$ and approximate this integral by a sequence of Riemann sums which converges in $C^\infty(X)$. We may take for instance

$$u_\varepsilon(x) = \left(\frac{\varepsilon}{2\pi}\right)^n \sum_{\nu \in (\varepsilon\mathbb{Z})^n, |\nu| \leq \frac{1}{\varepsilon}} e^{ix\cdot\nu} \hat{u}(\nu).$$

Since $A : C^\infty(X) \to C^\infty(X)$ is continuous we get

$$Au(x) = \int A(e^{ix\xi}) \hat{u}(\xi) \frac{d\xi}{(2\pi)^n} = \int e^{ix\xi} b(x,\xi) \hat{u}(\xi) \frac{d\xi}{(2\pi)^n},$$

with convergence in $C^\infty(X)$. $\square$

If $A \in L_{\rho,\delta}^m(X)$ is properly supported, then A belongs to $L^{-\infty}$ if and only if $\sigma_A \in S^{-\infty}$. If $A \in L_{\rho,\delta}^m(X)$ we can decompose $A = A' + A''$, where A' is properly supported and $A'' \in L^{-\infty}$, and we define σ_A in $S_{\rho,\delta}^m(X \times \mathbb{R}^n)/S^{-\infty}$ as the class of $\sigma_{A'}$. In this way we obtain a *bijective* map $L_{\rho,\delta}^m(X)/L^{-\infty} \to S_{\rho,\delta}^m(X \times \mathbb{R}^n)/S^{-\infty}$.

We next discuss *adjoints*. Let $(u \mid v) = \int u(x)\overline{v(x)}\,dx$ be the standard inner product on $L^2(X)$. If $A : C_0^\infty(X) \to \mathcal{D}(X)$ is continuous we define the (complex) adjoint $A^* : C_0^\infty(X) \to \mathcal{D}'(X)$ by $(Au \mid v) = (u \mid A^*v)$, $u, v \in C_0^\infty(X)$. Then for the distribution kernels, we have the relation

$$K_{A^*}(x,y) = \overline{K_A(y,x)}.$$

Theorem 3.5 Let $A \in L_{\rho,\delta}^m(X)$, $\rho > \delta$. Then $A^* \in L_{\rho,\delta}^m$ and $\sigma_{A^*}(x,\xi) \sim \sum \frac{1}{\alpha!} \partial_\xi^\alpha D_x^\alpha \overline{\sigma_A(x,\xi)}$.

Proof: If A is given by (3.1), then

$$A^*u(x) = \iint a^*(x,y,\xi) e^{i(x-y)\xi} u(y) \frac{dy\,d\xi}{(2\pi)^n},$$

where $a^*(x,y,\xi) = \overline{a(y,x,\xi)}$. Hence $A^* \in L_{\rho,\delta}^m$.

In order to compute σ_{A^*} we may assume that A is properly supported and take $a(x,y,\xi) = a(x,\xi) = \sigma_A(x,\xi)$ in (3.1). Then the preceding theorem tells us that

$$\sigma_{A^*}(x,\xi) = \sum \frac{1}{i^{|\alpha|}\alpha!} \left(\partial_\xi^\alpha \partial_y^\alpha \overline{(a(y,\xi))}\right)_{y=x},$$

which is the desired expansion. $\square$

We next consider the composition of pseudodifferential operators.

Theorem 3.6 Let $A \in L_{\rho,\delta}^{m'}(X)$, $B \in L_{\rho,\delta}^{m''}(X)$, $\rho > \delta$, with at least one of A, B properly supported. Then $A B \in L_{\rho,\delta}^{m'+m''}(X)$ and

$$\sigma_{A \circ B}(x, \xi) \sim \sum \frac{\partial_\xi^\alpha \sigma_A(x,\xi) \, D_x^\alpha \sigma_B(x,\xi)}{\alpha!} \tag{3.4}$$

Proof: After modifying A by an element of $L^{-\infty}$, which will modify $A \circ B$ by an operator which is continuous $\mathcal{E}' \to C^\infty$ and hence in $L^{-\infty}(X)$, we may assume that

$$Au(x) = \frac{1}{(2\pi)^n} \iint e^{i(x-y)\xi} a(x,\xi) \chi(x,y) u(y) \, dy \, d\xi$$

with χ as above and with $a(x,\xi) \sim \sigma_A(x,\xi)$.

We may also assume that

$$Bu(x) = \frac{1}{(2\pi)^n} \int e^{ix\xi} b(x,\xi) \hat{u}(\xi) \, d\xi, \quad u \in C_0^\infty(X),$$

where the integral can be approximated by a sequence of Riemann sums converging in $C^\infty(X)$, and $b(x,\xi) \sim \sigma_B(x,\xi)$.

Then for $u \in C_0^\infty(X)$, we obtain

$$A \circ Bu(x) = \frac{1}{(2\pi)^n} \int e^{ix\xi} c(x,\xi) \hat{u}(\xi) \, d\xi,$$

where $c(x,\xi) = e^{ix\xi} A(b(\cdot, \xi) e^{i(\cdot)\xi})$. As in the proof of Theorem 3.4 we find

$$c \in S_{\rho,\delta}^{m'+m''}, \quad c \sim \sum \frac{1}{\alpha!} \partial_\xi^\alpha a(x,\xi) \, D_x^\alpha b(x,\xi).$$

□

Remark 3.7 If $a \in S_{\rho,\delta}^{m'}(X \times \mathbb{R}^n)$, $b \in S_{\rho,\delta}^{m''}(X \times \mathbb{R}^n)$, $\rho > \delta$, then we can define $a \# b \in S_{\rho,\delta}^{m'+m''}(X \times \mathbb{R}^n)$ uniquely up to some element of $S^{-\infty}(X \times \mathbb{R}^n)$ by

$$(a \# b)(x,\xi) \sim \sum \frac{1}{\alpha!} \partial_\xi^\alpha a(x,\xi) \, D_x^\alpha b(x,\xi).$$

This gives a bilinear map (which we also denote by $\#$)

$$S_{\rho,\delta}^{m'}/S^{-\infty} \times S_{\rho,\delta}^{m''}/S^{-\infty} \ni (a, b) \mapsto a \# b \in S_{\rho,\delta}^{m'+m''}/S^{-\infty},$$

and it follows from Theorem 3.6 (this can also be verified directly (exercise)) that the "product" $\#$ is associative: $(a \# b) \# c = a \# (b \# c)$.

We finally discuss changes of variables and the corresponding notion of principal symbol. Let $\kappa : X \to \tilde{X}$ be a C^∞ map, where X, $\tilde{X}$ are open sets in $\mathbb{R}^n$, $\mathbb{R}^{\tilde{n}}$. Then if $u \in C^\infty(\tilde{X})$ we define $\kappa^* u \in C^\infty(X)$ by $\kappa^* u = u \circ \kappa$. If $d\kappa$ is surjective everywhere, then κ^* admits a unique continuous extension $\mathcal{D}'(\tilde{X}) \to \mathcal{D}'(X)$. We now assume that $\tilde{n} = n$ and that κ is a diffeomorphism (i.e. κ is bijective and the inverse κ^{-1} is a C^∞ map). If $\tilde{A}$ is a pseudodifferential operator on $\tilde{X}$, we want to study $A = \kappa^* \circ \tilde{A} \circ (\kappa^*)^{-1} : C_0^\infty(X) \to C^\infty(X)$, $\mathcal{E}'(X) \to \mathcal{D}'(X)$. For $u \in C_0^\infty(X)$ we find (if $\tilde{A}$ is given by (3.1) with a symbol $\tilde{a}(\tilde{x}, \tilde{y}, \tilde{\theta})$)

$$Au(x) = \tilde{A}(u \circ \kappa^{-1})(\kappa(x)) =$$
$$= \frac{1}{(2\pi)^n} \iint e^{i(\kappa(x) - \tilde{y})\tilde{\theta}} \tilde{a}(\kappa(x), \tilde{y}, \tilde{\theta}) u(\kappa^{-1}(\tilde{y})) \, d\tilde{y} \, d\tilde{\theta}.$$

After the change of variables $y = \kappa^{-1}(\tilde{y})$, we find

$$Au(x) = \frac{1}{(2\pi)^n} \iint e^{i(\kappa(x) - \kappa(y))\tilde{\theta}} \tilde{a}(\kappa(x), \kappa(y), \tilde{\theta}) \left|\det \frac{d\tilde{y}}{dy}\right| u(y) \, dy \, d\tilde{\theta}.$$

We assume $\tilde{a} \in S^m_{\rho, \delta}(\tilde{X} \times \tilde{X} \times \mathbb{R}^n)$, so that

$$\tilde{b}(x, y, \tilde{\theta}) \stackrel{\text{def}}{=} \tilde{a}(\kappa(x), \kappa(y), \tilde{\theta}) \left|\det \frac{d\tilde{y}}{dy}\right| \in S^m_{\rho, \delta}(X \times X \times \mathbb{R}^n).$$

Consider now the phase $\Phi(x, y, \tilde{\theta}) = (\kappa(x) - \kappa(y)) \tilde{\theta}$. By a change of variables in $\tilde{\theta}$ we want to obtain the standard phase for pseudodifferential operators. We already know that the distribution kernel K_A is of class C^∞ outside $\Delta(X \times X)$, so we restrict our attention to some suitable small neighborhood Ω of $\Delta(X \times X)$ in $X \times X$. We may assume that for every compact $K \subset \Omega$ there exists a compact $\tilde{K} \subset X$ such that $(x, y) \in K$, $t \in [0, 1] \Longrightarrow tx + (1-t)y \in \tilde{K}$.

Then for $(x, y) \in \Omega$ we write $\Phi(x, y, \tilde{\theta}) = \langle F(x, y)(x - y), \tilde{\theta} \rangle = \langle (x - y), G(x, y) \tilde{\theta} \rangle$, $\left(\langle a, b \rangle = a \cdot b = \sum_1^n a_j b_j \right)$,

$$G(x, y) = {}^t F(x, y), \quad F(x, y) = \int_0^1 \frac{\partial \kappa(tx + (1-t)y)}{\partial x} \, dt,$$

so that F and G are smooth in Ω and $F(x, x) = \frac{\partial \kappa}{\partial x}(x)$.

Possibly after decreasing Ω, we may assume that $G(x, y)$ is invertible for $(x, y) \in \Omega$. Then for $(x, y) \in \Omega$ we can make the change of variables and we get on Ω :

$$K_A(x, y) = \frac{1}{(2\pi)^n} \int e^{i(x-y)\theta} a(x, y, \theta) \, d\theta,$$

where $a(x,y,\theta) = \tilde{a}(\kappa(x), \kappa(y), G^{-1}(x,y)\theta) \dfrac{|\det \frac{\partial \kappa(y)}{\partial y}|}{|\det G(x,y)|}$. To determine the type of this new symbol, we see that x,y-derivatives of a will involve $\tilde{\theta}$-derivatives of $\tilde{a}$. For this reason we are led to assume that $\rho + \delta = 1$, and under that assumption we check that $a \in S^m_{\rho,\delta}(\Omega \times \mathbb{R}^n)$. We conclude that $A \in L^m_{\rho,\delta}(X)$. If we make the further hypothesis that $\rho > \delta$ (i.e. $\rho > \dfrac{1}{2}$ since $\delta = 1 - \rho$), we can also compute

$$(3.5) \quad \sigma_A(x,\xi) \sim \sum \frac{1}{\alpha!} \partial_\theta^\alpha D_y^\alpha \left(\sigma_{\tilde{A}}(\kappa(x), G(x,y)^{-1}\theta) \frac{|\det \frac{\partial \kappa(y)}{\partial y}|}{|\det G(x,y)|} \right) \bigg|_{y=x}$$

It is possible to show (also) by a direct argument, that $\sigma_A(x,\xi)$ does not depend on the choice of F in the expression for Φ. Here we shall only study the leading part. Put $L^m_\rho = L^m_{\rho,1-\rho}$, $S^m_\rho = S^m_{\rho,1-\rho}$.

Definition 3.8 If $A \in L^m_\rho(X)$, $\rho > \dfrac{1}{2}$ we define the principal symbol of A as the image of σ_A in $(S^m_\rho / S^{m-(2\rho-1)}_\rho)(X \times \mathbb{R}^n)$.

We then have a surjective map $L^m_\rho(X) \to (S^m_\rho / S^{m-(2\rho-1)}_\rho)(X \times \mathbb{R}^n)$ which gives rise to a bijection

$$L^m_\rho / L^{m-(2\rho-1)}_\rho \longrightarrow S^m_\rho / S^{m-(2\rho-1)}_\rho$$

Let $S^m_{\text{cl}}(X \times \mathbb{R}^N) \subset S^m_1(X \times \mathbb{R}^N)$ be the space of symbols $a(x,\theta)$ with

$$(3.6) \quad a(x,\theta) \sim \sum_{j=0}^\infty (1 - \chi(\theta)) a_{m-j}(x,\theta), \quad \chi \in C_0^\infty(\mathbb{R}^N), \chi = 1 \text{ near } 0,$$

with $a_k(x,\theta) \in C^\infty(X \times \dot{\mathbb{R}}^N)$ positively homogeneous of degree k in θ. Let $L^m_{\text{cl}}(X) \subset L^m_1(X)$ be the space of pseudodifferential operators A with $\sigma_A \in S^m_{\text{cl}}(X \times \mathbb{R}^n)$. For such an operator we can identify the principal symbol in $S^m_1(X \times \mathbb{R}^n)/S^{m-1}_1(X \times \mathbb{R}^n)$ with the positively homogenous function $a_m(x,\xi)$ if $a = \sigma_A$ has the asymptotic expansion (3.4).

Returning to the change of variables for $\tilde{A} \in L^m_\rho(\tilde{X})$, $\rho > \dfrac{1}{2}$, we see from (3.5) that

$$(3.7) \qquad \sigma_A(x,\xi) \equiv \sigma_{\tilde{A}}(\kappa(x), ({}^t\kappa'(x))^{-1}\xi) \mod S^{m-(2\rho-1)}_\rho$$

so if a, $\tilde{a}$ denote the principal symbols of A, $\tilde{A}$, we get the relation

$$(3.8) \qquad\qquad a(x, {}^t\kappa'(x)\tilde{\xi}) = \tilde{a}(\kappa(x), \tilde{\xi}).$$

Theorem 3.9 Let $\kappa : X \to \tilde{X}$ be a C^∞ diffeomorphism between two open sets in $\mathbb{R}^n$ and let $\tilde{A} \in L_\rho^m(\tilde{X})$; $\rho > \frac{1}{2}$. Then $A = \kappa^* \circ \tilde{A} \circ \kappa^{*-1}$ belongs to $L_\rho^m(X)$ and we have the relations (3.5) between the full symbols and (3.8) between the principal symbols.

Exercises

Exercise 3.1 (Weyl quantization)

Let $a \in \mathcal{S}(\mathbb{R}^{2n})$, $u \in \mathcal{S}(\mathbb{R}^n)$. Set

$$Op(a)\,u(x) = \iint e^{i(x-y)\theta} a\left(\frac{x+y}{2}, \theta\right) u(y)\, dy \, \frac{d\theta}{(2\pi)^n}.$$

a) Show that

$$[D_{x_j}, Op(a)] = Op\left(\frac{1}{i}\frac{\partial a}{\partial x_j}\right)$$

$$[x_j, Op(a)] = Op\left(-\frac{1}{i}\frac{\partial a}{\partial \theta_j}\right)$$

where in general we let $[M, N] = MN - NM$ denote the commutator of the operators M, N.

b) Let $S_{0,0}^0 = \{a \in C^\infty(\mathbb{R}^{2n})\,;\, \partial_x^\alpha \partial_\theta^\beta a \in L^\infty \;\; \forall (\alpha, \beta) \in \mathbb{N}^{2n}\}$.

Show that for $a \in \mathcal{S}(\mathbb{R}^{2n})$, $Op(a) : \mathcal{S}(\mathbb{R}^n) \to L^\infty(\mathbb{R}^n)$ is continuous and the continuity is uniform for $a \in \mathcal{S}(\mathbb{R}^{2n}) \cap B$ if B is a bounded set in $S_{0,0}^0$.

c) Show the same result for $x^\alpha \partial_x^\beta Op(a)$ and show that $Op(a) : \mathcal{S}(\mathbb{R}^n) \to \mathcal{S}(\mathbb{R}^n)$ is uniformly continuous for $a \in \mathcal{S}(\mathbb{R}^{2n}) \cap B$.

d) Let $a \in S_{0,0}^0$ and $\chi \in \mathcal{S}(\mathbb{R}^{2n})$, $\chi(0) = 1$. Show that $\lim_{\varepsilon \to 0} Op(\chi(\varepsilon(\cdot,\cdot))a)\,u$ exists for all $u \in \mathcal{S}(\mathbb{R}^n)$ and defines a linear continuous operator $Op(a) : \mathcal{S}(\mathbb{R}^n) \to \mathcal{S}(\mathbb{R}^n)$.

e) If $a \in S_{0,0}^0$, show that $Op(a)$ can be extended to a continuous operator $\mathcal{S}'(\mathbb{R}^n) \to \mathcal{S}'(\mathbb{R}^n)$.

f) Calculate $Op(a)$ in the following cases

$$a(x,\theta) = e^{i\ell_x \cdot x} \quad a(x,\theta) = e^{i\ell_\theta \cdot \theta} \quad a(x,\theta) = e^{i(\ell_x \cdot x + \ell_\theta \cdot \theta)}$$

where $\ell_\theta \in \mathbb{R}^n$, $\ell_x \in \mathbb{R}^n$.

g) For $n = 1$, $a \in C^\infty(\mathbb{R}^2)$, 2π-periodic in (x, θ), let

$$\hat{a}_{j,k} = (2\pi)^{-2} \iint e^{-i(jx+k\theta)} a(x, \theta) \, dx \, d\theta, \quad j, k \in \mathbb{Z},$$

be the Fourier coefficients of a. Show that $Op(a) : L^2 \to L^2$ is bounded and that

$$\|Op(a)\| \le \sum_{j,k} |\hat{a}_{j,k}|.$$

h) For $a \in \mathcal{S}(\mathbb{R}^2)$ (or even for $a \in S^0_{0,0}$ with $\hat{a} \in L^1$) show that $Op(a)$ is bounded $L^2 \to L^2$ and

$$\|Op(a)\| \le \frac{1}{(2\pi)^2} \|\hat{a}\|_{L^1}.$$

i) Show that for $a \in \mathcal{S}(\mathbb{R}^{2n})$, and $\mathcal{F}$ denoting the Fourier transform,

$$\mathcal{F}^{-1} Op(a) \mathcal{F} = Op(b)$$

where $b(x, \xi) = a \circ \kappa_\mathcal{F}(x, \xi)$, $\kappa_\mathcal{F}(x, \xi) = (\xi, -x)$.

j) Let $a, b \in \mathcal{S}(\mathbb{R}^{2n})$. Show that

$$Op(a) \, Op(b) = Op(c)$$

with

$$c(x, \xi) = \left(\exp \frac{i}{2} \sigma(D_x, D_\xi ; D_y, D_\eta) a(x, \xi) b(y, \eta) \right)\Big|_{\substack{y=x \\ \eta=\xi}}.$$

Here $\sigma(x, \xi; y, \eta) = \xi \cdot y - x \cdot \eta$.

Exercise 3.2

Let W be a neighborhood of $0 \in \mathbb{R}^n$ and $f \in C^\infty(W, \mathbb{R}^n)$ such that

$$f(x) = x - h(x), \quad h(x) = \mathcal{O}(|x|^2).$$

Let $j(f)$ be the Jacobian of f.

a) Show that f is a C^∞ diffeomorphism $V \to f(V)$, where V is a neighborhood of 0.
Show that

$$f^{-1}(x) = x - k(x), \quad k(x) = \mathcal{O}(|x|^2).$$

b) Let $T : C_0^\infty(V) \to C^\infty(f(V))$ be defined by $Tu(x) = u(f^{-1}(x))$. Show that T is a Fourier integral operator

$$Tu(x) = \frac{1}{(2\pi)^n} \iint e^{i(x-y)\xi} e^{ih(y)\xi} j(f)(y) u(y) \, dy \, d\xi, \quad u \in C_0^\infty(V).$$

c) Let $\partial_x^\alpha = \partial_{x_1}^{\alpha_1} \ldots \partial_{x_n}^{\alpha_n}$ and $h_\alpha(x) = h_1^{\alpha_1}(x) \ldots h_n^{\alpha_n}(x)$.
Show that for every N:

$$Tu(x) = \sum_{|\alpha|\leq N} \frac{\partial_x^\alpha}{\alpha!}\left(h^\alpha(x)\, j(f)(x)\, u(x)\right) + R_N(x),$$

$$R_N(x) =$$
$$\frac{1}{(2\pi)^n}\int_0^1 \iint e^{i(x-y)\xi}e^{ith(y)\xi}\frac{(ih(y)\cdot\xi)^{N+1}}{N!}(1-t)^N j(f)(y)\, u(y)\, dy\, d\xi\, dt.$$

d) For $t \in [0,1]$ let f_t be defined by $f_t(x) = x - th(x)$.
Show that if V is sufficiently small it is possible to write

$$R_N(x) = \int_0^1 (1-t)^N \sum_{|\alpha|=N+1} \frac{\partial_x^\alpha}{\alpha!} v_\alpha(f_t^{-1}(x), t)\, dt$$

where $v_\alpha(y,t) = h^\alpha(y)\dfrac{j(f)(y)}{j(f_t)(y)}u(y)$.
Show that $R_N(x) = \mathcal{O}(|x|^{N+1})$.

e) Derive the following identity between formal series (u is here viewed as a formal series)

$$u(f^{-1}(x)) = \sum_\alpha \frac{\partial_x^\alpha}{\alpha!}\left(h^\alpha(x)\, j(f)(x)\, u(x)\right)$$

(Abhyankhar's formula).
Get in particular for $u = x_k$

$$(f^{-1})_k(x) = \sum_\alpha \frac{\partial_x^\alpha}{\alpha!}\left(h^\alpha(x)\, j(f)(x)\, x_k\right).$$

Exercise 3.3 (Distributions on a manifold)

Let M be a compact C^∞ manifold and $\kappa_j : M_j \to X_j$, $j = 1, \ldots, N$ a set of local charts.

($M = \cup M_j$, M_j open, X_j open subset of $\mathbb{R}^n$ and κ_j is a diffeomorphism.)

A *distribution*, on M, $u \in \mathcal{D}'(M)$ is defined by a set of representatives $u_j \in \mathcal{D}'(X_j)$, $j = 1, \ldots, N$ with

$$u_j \circ (\kappa_j \circ \kappa_k^{-1}) = u_k \quad \text{in} \quad \kappa_k(M_j \cap M_k).$$

Show that this definition does not depend on the choice of the set of local charts and show how to define the new representatives of u in another set of local charts.

Exercise 3.4 (Pseudodifferential operators on a manifold)

Let M, M_j, κ_j be as in Exercise 3.3. By definition, a linear continuous operator $P : C^\infty(M) \to \mathcal{D}'(M)$ belongs to $L^m_{\text{cl}}(M)$ if for all $j \in \{1,\ldots,N\}$ and $u \in C_0^\infty(M_j)$ we have $Pu \circ \kappa_j^{-1} = P_j(u \circ \kappa_j^{-1})$ with $P_j \in L^m_{\text{cl}}(X_j)$, and $\chi_1 P \chi_2 : \mathcal{D}'(M) \to C^\infty(M)$ for all $\chi_1, \chi_2 \in C^\infty(M)$ with $\operatorname{supp}\chi_1 \cap \operatorname{supp}\chi_2 = \emptyset$.

Let $P \in L^m_{\text{cl}}(M)$, $Q \in L^k_{\text{cl}}(M)$.

1) Show that $P : C^\infty(M) \to C^\infty(M)$ and extends to a continuous linear map $\mathcal{D}'(M) \to \mathcal{D}'(M)$.

2) Show that $P \circ Q \in L^{m+k}_{\text{cl}}(M)$.

Let ω be a strictly positive smooth density on M, so that $\int_M u\omega$ is well defined for every $u \in \mathcal{D}'(M)$. (This means that there are functions $0 < f_j \in C^\infty(X_j)$ such that $\int_M u\omega = \sum_{j=1}^N \int (\chi_j u \circ \kappa_j^{-1})(x) f_j(x)\, dx$, when $1 = \sum_j \chi_j$, $\chi_j \in C_0^\infty(M_j)$. Write down the compatibility conditions for the f_j !) We then put $(u \mid v) = \int u(x)\overline{v(x)}\omega$ for $u,v \in C^\infty(M)$ and extend to a sesquilinear form on $C^\infty \times \mathcal{D}'$, $\mathcal{D}' \times C^\infty$, $L^2 \times L^2$. For $P \in L^m_{\text{cl}}(M)$ define $P^* : C^\infty(M) \to \mathcal{D}'(M)$ by $(Pu \mid v) = (u \mid P^* v)$, $u,v \in C^\infty(M)$.

3) Show that $P^* \in L^m_{\text{cl}}$.

4) (The principal symbol) Let p_j be the principal symbol of P_j (in the definition of P). Show that $p_k(\kappa_k \circ \kappa_j^{-1}(x), \xi) = p_j(x, {}^t d(\kappa_k \circ \kappa_j^{-1})(x)(\xi))$, $x \in \kappa_j(M_j \cap M_k)$, $\xi \in \mathbb{R}^n$, and define a naturally associated function $p \in C^\infty(T^*M \setminus 0)$, which is positively homogeneous of degree m. (See Chapter 5 for a review of the definition of T^*M.)

Conversely, if $p_j \in C^\infty(X_j \times \dot{\mathbb{R}}^n)$ are positively homogeneous of degree m and satisfy the compatibility conditions above, show that there exists $P \in L^m_{\text{cl}}(M)$ unique modulo $L^{m-1}_{\text{cl}}(M)$ such that P_j has the prinicpal symbol p_j.

5) Let $\rho > \frac{1}{2}$. Extend the definitions and the results of 1) – 4) to the class $L^m_\rho(M)$.

Notes

The presentation of this chapter follows that of [Hö2], but we have emphasized the role of the method of stationary phase by basing the proofs of Theorems 3.4 – 3.6 on Chapter 2. Pseudodifferential calculus with symbols of class S^m_{cl} was introduced by Kohn–Nirenberg [KN] and the symbol classes $S^m_{\rho,\delta}$ were introduced by Hörmander see [Hö2]. Subsequently many other symbol classes have been introduced, in particular by Beals–Fefferman [BF], [B], Hörmander [Hö3,4].

The change of fiber variables in the proof of Theorem 3.9 is called the Kuranishi trick (see pages 102, 107 in [Hö2]), and similar changes of variables are often practically useful. (See Chapter 10.)

4 Application to elliptic operators and L^2 continuity

Let $P \in L^m_{\rho,\delta}(X)$ and denote for simplicity its symbol by $P(x,\xi)$. (The corresponding operator will sometimes be denoted $P(x,D_x)$.) We say that P is elliptic or non-characteristic at the point $(x_0,\xi_0) \in X \times \dot{\mathbb{R}}^n$ if there exists a conical neighborhood V of (x_0,ξ_0) and a constant $C > 0$ such that $|P(x,\xi)| \geq \frac{1}{C}(1+|\xi|)^m$ for $(x,\xi) \in V$, $|\xi| \geq C$. When $P \in L^m_{cl}(X)$ has the homogenous principal symbol $p(x,\xi)$, then P is non-characteristic at (x_0,ξ_0) if and only if $p(x_0,\xi_0) \neq 0$.

We say that P is elliptic at $x_0 \in X$ if P is elliptic at (x_0,ξ_0) for every $\xi_0 \in \dot{\mathbb{R}}^n$ (or equivalently for every $\xi_0 \in S^{n-1}$, where $S^{n-1} = \{\xi \in \mathbb{R}^n; |\xi| = 1\}$). We say that P is elliptic on $Y \subset X$ if P is elliptic at every point x_0 in Y. We say that P is elliptic if P is elliptic on X.

Theorem 4.1 *If $P \in L^m_{\rho,\delta}(X)$ is elliptic and $\rho > \delta$, then there exists $Q \in L^{-m}_{\rho,\delta}(X)$, properly supported, such that $P \circ Q \equiv Q \circ P \equiv I \bmod L^{-\infty}(X)$. (I is the identity operator: $Iu = u$.) Moreover, Q is unique modulo $L^{-\infty}(X)$.*

Proof: Using a partition of unity we can first find a function $Q_0 \in C^\infty(X \times \mathbb{R}^n)$ such that for every compact $K \subset X$, there is a constant $C_K > 0$ such that
$$Q_0(x,\xi) = \frac{1}{P(x,\xi)} \quad \text{for} \quad x \in K, \ |\xi| \geq C_K.$$
□

Lemma 4.2 $Q_0(x,\xi) \in S^{-m}_{\rho,\delta}(X \times \mathbb{R}^n)$.

Proof: If suffices to estimate Q_0 and its derivatives in the regions $x \in K$, $|\xi| \geq C_K$. In such a region (K fixed), we first have $|Q_0(x,\xi)| \leq C(1+|\xi|)^{-m}$. By induction, we assume
$$|\partial_x^\alpha \partial_\xi^\beta Q_0(x,\xi)| \leq C_{\alpha,\beta}(1+|\xi|)^{-m-\rho|\beta|+\delta|\alpha|} \quad \text{for} \quad |\alpha|+|\beta| < N.$$
Let $|\alpha|+|\beta| = N$. Differentiating the relation $Q_0 \cdot P = 1$, we get
$$P \partial_x^\alpha \partial_\xi^\beta Q_0 = \sum_{\substack{\alpha'+\alpha''=\alpha \\ \beta'+\beta''=\beta \\ |\alpha''+\beta''|<N}} C_{\alpha',\beta',\alpha'',\beta''}(\partial_x^{\alpha'}\partial_\xi^{\beta'} P)(\partial_x^{\alpha''}\partial_\xi^{\beta''} Q_0)$$

and by the induction hypothesis, we can estimate the modulus of this expression by
$$\text{const.}\ (1+|\xi|)^{-\rho|\beta|+\delta|\alpha|}.$$

Hence

$$|\partial_x^\alpha \partial_\xi^\beta Q_0| \leq C_{\alpha,\beta}(1+|\xi|)^{-m-\rho|\beta|+\delta|\alpha|} \quad \text{for} \quad |\alpha|+|\beta| = N.$$

□

In the remainder of the proof we work in the symbol classes $S_{\rho,\delta}^m(X \times \mathbb{R}^n)/S^{-\infty}$. We have (cf. Remark 3.7) $P\#Q_0 = 1-R$, $Q_0\#P = 1-T$ with $R, T \in S_{\rho,\delta}^{-(\rho-\delta)}/S^{-\infty}$. Define

$$Q_r = Q_0\#(1 + R + R\#R + R\#R\#R + \ldots) \in S_{\rho,\delta}^m/S^{-\infty}$$
$$Q_\ell = (1 + T + T\#T + T\#T\#T + \ldots)\#Q_0 \in S_{\rho,\delta}^m/S^{-\infty}.$$

Then $P\#Q_r = 1$, $Q_\ell\#P = 1$ (in $S_{\rho,\delta}^0/S^{-\infty}$).
Moreover $Q_\ell = Q_\ell\#(P\#Q_r) = (Q_\ell\#P)\#Q_r = Q_r$.

Let $Q \in L_{\rho,\delta}^m(X)$ be properly supported with symbol $Q_\ell = Q_r \mod S^{-\infty}$. Then (on the level of operators) $P \circ Q \equiv Q \circ P \equiv I \mod L^{-\infty}(X)$. If $Q' \in L_{\rho,\delta}^m(X)$ is a second operator with the same properties, we get $P \circ (Q-Q') \equiv 0$ and composing with Q to the left, we get $Q - Q' \equiv 0$.

□

The operator Q in Theorem 4.1 is called a parametrix.

Corollary 4.3 Let $\rho > \delta$. If $P \in L_{\rho,\delta}^m(X)$ is elliptic and properly supported, then P induces a bijection

$$\mathcal{D}'(X)/C^\infty(X) \longrightarrow \mathcal{D}'(X)/C^\infty(X).$$

Corollary 4.4 Under the same assumptions on P, we have $\operatorname{sing\,supp}(Pu) = \operatorname{sing\,supp}(u)$ for every $u \in \mathcal{D}'(X)$.

We now consider the action of pseudodifferential operators in Sobolev spaces. The basic result for this will be the L^2 continuity of operators of order 0 and type ρ, δ with $\rho > \delta$.

(The Calderón–Vaillancourt theorem states that we still have L^2 continuity when $\rho = \delta$; see Exercise 4.7 for the crucial case $\rho = \delta = 0$.)

Theorem 4.5 Let $A \in L_{\rho,\delta}^0(\mathbb{R}^n)$, $\rho > \delta$ have compactly supported distribution kernel: $\operatorname{supp} K_A \in \mathcal{E}'(\mathbb{R}^n \times \mathbb{R}^n)$. Let $M = \limsup_{|\xi| \to \infty} \sup_{x \in \mathbb{R}^n} |\sigma_A(x, \xi)|$. Then A is continuous $L^2(\mathbb{R}^n) \to L^2(\mathbb{R}^n)$ and for every $\varepsilon > 0$, we can find a decomposition $A = A_\varepsilon + K_\varepsilon$, where $\|A_\varepsilon\| \leq M + \varepsilon$, $K_\varepsilon \in L^{-\infty}(\mathbb{R}^n)$, $\operatorname{supp} K_{K_\varepsilon} \subset (\operatorname{supp} K_A) + \{0\} \times B(0, \varepsilon)$. Here $\|\ \|$ denotes the norm for bounded operators on $L^2(\mathbb{R}^n)$ and $B(0, \varepsilon) \subset \mathbb{R}^n$ is the open ball of center 0 and radius ε.

Proof: The idea of the proof is to try to show that $A^*A \leq (M+\varepsilon)^2 I$ (in the sense of self-adjoint operators) by constructing a square root of $(M+\varepsilon)^2 I - A^*A$.

Lemma 4.6 *For every $\varepsilon > 0$, there exists $B \in L^0_{\rho,\delta}(\mathbb{R}^n)$ with supp $K_{(M+\varepsilon)I-B} \in \mathcal{E}'(\mathbb{R}^n \times \mathbb{R}^n)$, such that $(M+\varepsilon)^2 I = A^*A + B^*B + K$, where $K \in L^{-\infty}(\mathbb{R}^n)$.*

Proof of the lemma : We consider the formally self-adjoint operator $C = (M+\varepsilon)^2 I - A^*A$. Then $|\sigma_C(x,\xi)| \geq$ const. for $|\xi|$ sufficiently large and $\sigma_C - \bar{\sigma}_C \in S^{-(\rho-\delta)}_{\rho,\delta}$. We can then find $b_0 \in S^0_{\rho,\delta}$ (with $b_0 = M + \varepsilon$ for x outside some compact set) such that $\sigma_C - \bar{b}_0 b_0 \in S^{-(\rho-\delta)}_{\rho,\delta}$. If $B_0 \in L^0_{\rho,\delta}(\mathbb{R}^n)$ is of the form $(M+\varepsilon) I$ + compactly supported operator and has the symbol $b_0 \bmod S^{-\infty}$, then $B_0^* B_0 = C + C_1$, where $C_1 \in L^{-(\rho-\delta)}_{\rho,\delta}$ is formally self-adjoint. If $B_0^{-1} \in L^0_{\rho,\delta}$ is a parametrix of B_0, we put $B_1 = B_0 - \frac{1}{2}(B_0^{-1})^* C_1$ and find

$$B_1^* B_1 = C + C_2, \quad \text{where} \quad C_2 \in L^{-2(\rho-\delta)}_{\rho,\delta}.$$

Iterating this argument we find $B \sim B_0 - \frac{1}{2}(B_0^{-1})^* C_1 - \frac{1}{2}(B_0^{-1})^* C_2 \ldots$. Inspection of the arguments shows that we can choose B with all the desired properties. □

For $u \in C_0^\infty(\mathbb{R}^n)$ we obtain, since $(B^*Bu \mid u) = \|Bu\|^2 \geq 0$, that $\|Au\|^2 \leq (M+\varepsilon)^2 \|u\|^2 + |(Ku \mid u)| \leq \text{const.}\|u\|^2$, where the norms and the scalar products are those of $L^2(\mathbb{R}^n)$. Hence A is a bounded operator on $L^2(\mathbb{R}^n)$.

Let $0 \leq \psi \in C_0^\infty(B(0,\frac{1}{2}))$ satisfy $\int \psi(x)\,dx = 1$. Then $0 \leq |\hat{\psi}| \leq 1$, $\hat{\psi}(0) = 1$. Put $\chi = \psi * \check{\psi} \in C_0^\infty(B(0,1))$, where $\check{\psi}(x) = \psi(-x)$ and $*$ indicates convolution $\left(u * v(x) = \int u(x-y)\,v(y)\,dy\right)$. Then $\hat{\chi} = |\hat{\psi}|^2$, so $0 \leq \hat{\chi} \leq 1$ and $\hat{\chi}(0) = 1$. Put $P_\varepsilon u = u - \chi_\varepsilon * u$, where $\chi_\varepsilon(x) = \varepsilon^{-n} \chi(\frac{x}{\varepsilon})$. By Fourier and Plancherel we have $\|P_\varepsilon u\| \leq \|u\|$. Applying this to the estimate above for $\|Au\|^2$, we get

$$\begin{aligned}\|AP_\varepsilon u\|^2 &\leq (M+\varepsilon)^2 \|u\|^2 + |(KP_\varepsilon u \mid P_\varepsilon u)| \\ &\leq (M+\varepsilon)^2 \|u\|^2 + \|KP_\varepsilon u\| \, \|u\|.\end{aligned}$$

The operator KP_ε has the distribution kernel

$$K_\varepsilon(x,y) = K(x,y) - \int K(x,y')\,\check{\chi}_\varepsilon(y-y')\,dy'$$

which has uniformly compact support and tends to zero uniformly when $\varepsilon \to 0$. It follows that the norm of KP_ε tends to 0 when ε tends to zero, and for any given $\varepsilon_0 > 0$ we find $\|AP_\varepsilon u\| \leq (M + \varepsilon_0)\|u\|$ if $\varepsilon > 0$ is sufficiently small. We have $Au = A_\varepsilon u + A(\chi_\varepsilon * u)$ with $A_\varepsilon = AP_\varepsilon$, supp $K_{A_\varepsilon} \subset$ supp $K_A + \{0\} \times B(0,\varepsilon)$, and this concludes the proof of the theorem. (The "ε" that we finally choose in the proof may be much smaller than the "ε" in the formulation of the theorem.) □

We can now relate pseudodifferential operators and the standard (L^2-based) Sobolev spaces $H^s(\mathbb{R}^n)$, $s \in \mathbb{R}$. Recall that $H^s(\mathbb{R}^n)$ is the space of $u \in \mathcal{S}'(\mathbb{R}^n)$ such that $\hat{u}(\xi)$ is locally square integrable and such that
$$\|u\|^2_{H^s(\mathbb{R}^n)} = \|u\|^2_s = \frac{1}{(2\pi)^n}\int |\hat{u}(\xi)|^2(1+|\xi|^2)^s\, d\xi < \infty.$$

For $s \in \mathbb{N}$ it is also the space of $u \in L^2(\mathbb{R}^n)$ such that $D^\alpha u \in L^2(\mathbb{R}^n)$ for $|\alpha| \leq s$.

We recall some of the properties of these spaces :

1) $H^s(\mathbb{R}^n)$ is a Banach space with the norm $\|u\|_s$.

2) The dual of H^s for the standard L^2 inner product $(u\,|\,v) = \int u(x)\,\overline{v(x)}\, dx$ is H^{-s}.

3) H^s is also a Hilbert space with the inner product
$$(u\,|\,v)_{H^s} = \frac{1}{(2\pi)^n}\int \hat{u}(\xi)\,\overline{\hat{v}(\xi)}\,(1+|\xi|^2)^s\, d\xi.$$

4) Every Hilbert space can be identified with its own dual. In our case the identification is given by $H^s(\mathbb{R}^n) \ni v \mapsto w \in H^{-s}(\mathbb{R}^n)$, with $\hat{w}(\xi) = (1+|\xi|^2)^s\,\hat{v}(\xi)$ or in pseudodifferential form $w = (1+|D|^2)^s\,v$.

5) More generally $(1+|D|^2)^{\ell/2}$ gives an isomorphism $H^s(\mathbb{R}^n) \to H^{s-\ell}(\mathbb{R}^n)$.

6) We have $H^s(\mathbb{R}^n) \subset H^t(\mathbb{R}^n)$ if $s > t$ and the inclusion map $H^s(\mathbb{R}^n) \cap \mathcal{E}'(K) \hookrightarrow H^t(\mathbb{R}^n)$ is compact for every compact $K \subset \mathbb{R}^n$. If $\varepsilon > 0$, then $C(\mathbb{R}^n) \cap L^\infty(\mathbb{R}^n) \supset H^{\frac{n}{2}+\varepsilon}(\mathbb{R}^n)$, $C^k(\mathbb{R}^n) \supset H^{\frac{n}{2}+k+\varepsilon}(\mathbb{R}^n)$, $k \in \mathbb{N}$.

7) $H^s(\mathbb{R}^n)$ is a local space : if $\varphi \in C_0^\infty(\mathbb{R}^n)$, $u \in H^s(\mathbb{R}^n)$, then $\varphi u \in H^s(\mathbb{R}^n)$ and the corresponding operator of multiplication is continuous : $\|\varphi u\|_s \leq C_\varphi \|u\|_s$, $\forall u \in H^s(\mathbb{R}^n)$. If $X \subset \mathbb{R}^n$ is open, we can therefore define $H^s_{\mathrm{loc}}(X) = \{u \in \mathcal{D}'(X)\,;\,\varphi u \in H^s(\mathbb{R}^n),\,\forall \varphi \in C_0^\infty(X)\}$. This is a Fréchet space and the corresponding dual space turns out to be $H^{-s}_{\mathrm{comp}}(X) = H^{-s}_{\mathrm{loc}}(X) \cap \mathcal{E}'(X)$.

Theorem 4.7 Let $A \in L^m_{\rho,\delta}(X)$, $\rho > \delta$, be properly supported. Then A is continuous : $H^s_{\mathrm{loc}}(X) \to H^{s-m}_{\mathrm{loc}}(X)$, $H^s_{\mathrm{comp}}(X) \to H^{s-m}_{\mathrm{comp}}(X)$ for every $s \in \mathbb{R}$.

If A is elliptic then for every $u \in \mathcal{D}'(X)$ we have $u \in H^s_{\text{loc}}(X)$ if and only if $Au \in H^{s-m}_{\text{loc}}(X)$.

Proof: To prove (for example) the continuity of $A : H^s_{\text{loc}}(X) \to H^{s-m}_{\text{loc}}(X)$, we have to show that for $\varphi \in C_0^\infty(X)$, φA is continuous $H^s_{\text{loc}}(X) \to H^{s-m}_{\text{loc}}(\mathbb{R}^n)$. For such a φ, let $\psi \in C_0^\infty(X)$ be equal to 1 near $C^{-1}(\operatorname{supp}\varphi)$, where $C = \operatorname{supp} K_A$ is viewed as a relation. Then $\varphi A = \varphi A \psi$ and if suffices to prove that $\tilde{A} \stackrel{\text{def}}{=} \varphi A \psi$ is continuous $H^s(\mathbb{R}^n) \to H^{s-m}(\mathbb{R}^n)$. (We systematically identify functions on X having compact support, with the corresponding extensions by zero to $\mathbb{R}^n$.) Here $\tilde{A} \in L^m_{\rho,\delta}(\mathbb{R}^n)$, $\operatorname{supp} K_{\tilde{A}} \in \mathcal{E}'(\mathbb{R}^{2n})$.

We next observe that $(1 + |D|^2)^{\ell/2} u = v_\ell * u$, where

$$v_\ell(x) = \frac{1}{(2\pi)^n} \int e^{ix\xi}(1+\xi^2)^{\ell/2}\, d\xi.$$

It is easy to see that v_ℓ is of class $\mathcal{S}$ outside 0 and we can therefore find $\chi_\ell \in C_0^\infty(\mathbb{R}^n)$ equal to 1 near 0 such that $w_\ell \stackrel{\text{def}}{=} \chi_\ell v_\ell$ everywhere satisfies $\hat{w}_\ell(\xi) - (1+\xi^2)^{\ell/2} \in \mathcal{S}$, $\hat{w}_\ell(\xi) \neq 0$. Then the operator Λ_ℓ of convolution with w_ℓ is properly supported and gives isomorphisms $H^s(\mathbb{R}^n) \to H^{s-\ell}(\mathbb{R}^n)$ for every $s \in \mathbb{R}$. To show the continuity of $\tilde{A} : H^s(\mathbb{R}^n) \to H^{s-m}(\mathbb{R}^n)$ it now suffices to show the continuity of $B \stackrel{\text{def}}{=} \Lambda_{s-m} \tilde{A} \Lambda_{-s} : H^0(\mathbb{R}^n) \to H^0(\mathbb{R}^n)$. Since $\Lambda_\ell \in L^\ell_{1,0}(\mathbb{R}^n)$ is properly supported we know that $B \in L^0_{\rho,\delta}(\mathbb{R}^n)$, $K_B \in \mathcal{E}'(\mathbb{R}^{2n})$ and the required continuity of B follows from Theorem 4.5. $\square$

Corollary 4.8 *(Local solvability.) Let A be an elliptic differential operator with smooth coefficients on an open set $X \subset \mathbb{R}^n$ and let $x_0 \in X$. Then there exists an open neighborhood $V \subset X$ of x_0 such that for every $v \in \mathcal{D}'(V)$ and every open $W \subset\subset V$, there exists $u \in \mathcal{D}'(V)$ such that $Au = v$ in W.*

Proof: We follow the classical method of a priori estimates. For every compact $K \subset X$, $s \in \mathbb{R}$, there exists $C = C_{K,s} > 0$ such that

$$\|u\|_{s+m} \leq C(\|A^*u\|_s + \|u\|_s), \quad u \in \mathcal{E}'(K) \cap H^{s+m}(\mathbb{R}^n).$$

In fact, let $B \in L^{-m}_{1,0}(X)$ be a properly supported parametrix of A^*. Then $u = BA^*u + Ku$ where $K \in L^{-\infty}(X)$ is properly supported and both B and K are continuous $H^s_{\text{comp}}(X) \to H^{s+m}_{\text{comp}}(X)$.

The case $m = 0$ being trivial, we assume $m \geq 1$. By the Poincaré lemma we know that for every $\varepsilon > 0$ we have $\|u\|_0 \leq \varepsilon \|u\|_m$ for $u \in \mathcal{E}'(\mathbb{R}^n) \cap H^m(\mathbb{R}^n)$ if the diameter of the support of u is sufficiently small depending on ε, m only. Hence, if V is a sufficiently small open neighborhood of x_0, we have in addition to the previous estimate with $s = 0$, that $C\|u\|_0 \leq \frac{1}{2}\|u\|_m$ and hence

$$\|u\|_m \leq 2C\|A^*u\|_0, \quad u \in \mathcal{E}'(V) \cap H^m(\mathbb{R}^n).$$

If $v \in \mathcal{D}'(V)$, we first put $\tilde{u} = \tilde{B}v$, where $\tilde{B} \in L_{1,0}^{-m}(V)$ is a properly supported parametrix of A. Then $A\tilde{u} = v + \tilde{v}$ where $\tilde{v} \in C^\infty(V)$, and the problem of local solvability is reduced to the case when $v \in C^\infty(V)$. For such a v, we let $W \subset\subset V$ be open and consider the linear form

$$\ell : H^m(\mathbb{R}^n) \cap \mathcal{E}'(W) \ni \varphi \mapsto (\varphi \mid v) \in \mathbb{C}.$$

Then $|\ell(\varphi)| \le C(v,W)\|\varphi\|_m \le \tilde{C}(v,W)\|A^*\varphi\|_0$. Hence $\ell(\varphi) = k(A^*\varphi)$, where k is a bounded linear form on $L = \{A^*\varphi \in H^0 \cap \mathcal{E}'(W)\,;\, \varphi \in H^m(\mathbb{R}^n) \cap \mathcal{E}'(W)\}$.

By the Hahn–Banach theorem, k has a bounded extension to $H^0(\mathbb{R}^n)$ so there exists $u \in H^0(\mathbb{R}^n)$ such that

$$k(A^*\varphi) = (A^*\varphi \mid u), \quad \forall \varphi \in H^m \cap \mathcal{E}'(W).$$

Then $Au = v$ in W. $\square$

Exercises

Exercise 4.1

Sobolev spaces
Let us recall the definition

$$H^s(\mathbb{R}^n) = \{u \in \mathcal{S}'(\mathbb{R}^n)\,;\, \langle\xi\rangle^s \hat{u}(\xi) \in L^2(\mathbb{R}^n)\}$$
$$\|u\|_{H^s}^2 = \frac{1}{(2\pi)^n} \|\langle\xi\rangle^s \hat{u}\|_{L^2}^2$$

(with $\langle\xi\rangle^s = (1+\xi^2)^{s/2}$, $s \in \mathbb{R}$).

1) If $s \in \mathbb{N}$, show that $H^s(\mathbb{R}^n) = \{u \in L^2(\mathbb{R}^n)\,;\, D^\alpha u \in L^2(\mathbb{R}^n)\,|\alpha| \le s\}$ and that for some $C = C(n,s)$:

$$\frac{1}{C}\|u\|_{H^s}^2 \le \sum_{|\alpha|\le s} \|D^\alpha u\|_{L^2}^2 \le C\|u\|_{H^s}^2, \quad u \in H^s(\mathbb{R}^n).$$

2) Let $x = (x', x'')$ $x' \in \mathbb{R}^{n-d}$, $x'' \in \mathbb{R}^d$. If $s > \dfrac{d}{2}$ show that the operator

$$u \in \mathcal{S}(\mathbb{R}^n) \to u|_{x''=0} \in \mathcal{S}(\mathbb{R}^{n-d}_{x'})$$

can be extended to a bounded operator $H^s(\mathbb{R}^n) \to H^{s-d/2}(\mathbb{R}^{n-d})$.

3) If $s > \dfrac{n}{2}$ show that $\mathcal{S}(\mathbb{R}^n) \times \mathcal{S}(\mathbb{R}^n) \ni (u,v) \mapsto uv \in \mathcal{S}(\mathbb{R}^n)$ can be extended to a bilinear continuous map $H^s \times H^s \to H^s$.

4) If Ω is an open subset of $\mathbb{R}^n$ show that $H^s_{\text{loc}}(\Omega)$ is a Fréchet space.

5) Let $\kappa : W \to \Omega$ be a C^∞ diffeomorphism, W, Ω open subsets of $\mathbb{R}^n$. Show that if $u \in H^s_{\text{loc}}(\Omega)$ then $\kappa^*(u) \in H^s_{\text{loc}}(W)$. Define and study the spaces $H^s(M)$ when M is a compact C^∞ manifold in the spirit of Exercise 3.4.

6) Show that $\|u\|_0 \leq \varepsilon(\delta) \|u\|_1$, $u \in C^\infty_0(B(0,\delta))$ where $\varepsilon(\delta) \to 0$, $\delta \to 0$.
(Hint : consider $(u \mid \sum x_j \partial_{x_j} u)$.)

7) Show that the inclusion $H^s(\mathbb{R}^n) \cap \mathcal{E}'(K) \hookrightarrow H^{s'}(\mathbb{R}^n)$ is compact if K is compact and $s > s'$.
(Hint : let $\chi \in C^\infty_0(\mathbb{R}^n)$, $\chi = 1$ in a neighborhood of K and $u \in \mathcal{E}'(K)$. Then $\chi u = u$ and $\hat{u} = \dfrac{1}{(2\pi)^n} \hat{\chi} * \hat{u}$. Show that if $(u_j) \in H^s(\mathbb{R}^n) \cap \mathcal{E}'(K)$ is a bounded sequence, then there exists a subsequence u_{j_ν} converging in $H^{s'}(\mathbb{R}^n)$.)

Exercise 4.2

Let $0 < a \in C^\infty(\mathbb{R})$ and $p(x,\xi) = 1 + a(x_1)\xi_1^2 + i\xi_2$, $x = (x_1, x_2)$, $\xi = (\xi_1, \xi_2)$.

1) Show that for all $\alpha, \beta \in \mathbb{N}^2$ and every compact $K \subset \mathbb{R}^2$, there exists $C = C_{K,\alpha,\beta}$ such that
$$|\partial^\alpha_x \partial^\beta_\xi p(x,\xi)| \leq C|p(x,\xi)|^{1-\frac{1}{2}\beta_1-\beta_2}, \quad (x,\xi) \in K \times \mathbb{R}^2.$$

2) Let $q(x,\xi) = \dfrac{1}{p(x,\xi)}$. Show that for all $\alpha, \beta \in \mathbb{N}^2$ and every compact $K \subset \mathbb{R}^2$, there exists $C = C_{K,\alpha,\beta}$ such that
$$|\partial^\alpha_x \partial^\beta_\xi q(x,\xi)| \leq C|p(x,\xi)|^{-1-\frac{1}{2}\beta_1-\beta_2}, \quad (x,\xi) \in K \times \mathbb{R}^2.$$
Deduce that $q \in S^{-1}_{\frac{1}{2},0}(\mathbb{R}^2 \times \mathbb{R}^2)$.

Consider $P(x,D) = 1 - a_1(x)\partial^2_{x_1} + \partial_{x_2}$ on an open subset Ω of $\mathbb{R}^2$. Let $q(x,D) \in L^{-1}_{\frac{1}{2},0}(\Omega)$ be a properly supported operator with symbol $q(x,\xi)$.

3) Show that $q(x,D) \circ P(x,D) = I - R$, $R \in L^{-\frac{1}{2}}_{\frac{1}{2},0}(\Omega)$.

4) Show that there exists $Q \in L^{-1}_{\frac{1}{2},0}(\Omega)$ such that $Q(x,D) \circ P(x,D) = I - K$, $K \in L^{-\infty}(\Omega)$.

5) Let $u \in \mathcal{D}'(\Omega)$. What can be said about u if $Pu \in C^\infty(\Omega)$? What can be said if $Pu \in H^s_{\text{loc}}(\Omega)$?

Exercise 4.3

Let $x = (x', x_n) \in \mathbb{R}^n$, $\mathbb{R}^n_+ = \{x_n > 0\}$, $\overline{\mathbb{R}^n_+} = \{x_n \geq 0\}$.

1) Find a linear operator $K : \mathcal{S}(\mathbb{R}^{n-1}) \to C^\infty(\overline{\mathbb{R}^n_+})$ such that $u = Kw$ solves the Dirichlet problem

$$\begin{cases} \Delta u = 0 & \text{in } \mathbb{R}^n_+ \\ u|_{x_n=0} = w. \end{cases}$$

(Use the partial Fourier transform in the x' variable.)

2) Let $\ell_j \in C^\infty(\mathbb{R}^{n-1}, \mathbb{R})$, $j = 1, \ldots, n$. Show that

$$Qw = \left(\sum_{j=1}^n \ell_j(x') D_{x_j} Kw \right)\Big|_{x_n=0}$$

defines a pseudodifferential operator $Q \in L^1_{\text{cl}}(\mathbb{R}^{n-1})$ whose total symbol is

$$\sigma_Q(x', \xi') = \sum_{j=1}^{n-1} \ell_j(x') \xi'_j + i\,\ell_n(x') |\xi'|.$$

3) If $n = 2$, show that Q is elliptic if and only if $\vec{\ell} = (\ell_1, \ell_2) \neq 0$, $\forall x' \in \mathbb{R}^{n-1}$. If $n \geq 3$, show that Q is elliptic if and only if $\ell_n(x') \neq 0$, $\forall x' \in \mathbb{R}^{n-1}$.

Exercise 4.4

Let X be an open subset of $\mathbb{R}^n$, and let $P \in L^1_{\text{cl}}(X)$ be properly supported with principal symbol p, positively homogeneous of degree 1 in ξ. (See also Exercise 5.5.)

Assume $\frac{1}{i}\{p, \bar{p}\} < 0$, where in general $\{a, b\} = \sum_{1}^n \partial_{\xi_j} a\, \partial_{x_j} b - \partial_{x_j} a\, \partial_{\xi_j} b$ denotes the Poisson bracket.

1) Show that $\forall K \subset\subset X$, $\exists C_K > 0$ such that

$$\forall u \in C_0^\infty(X) \cap \mathcal{E}'(K) \quad \|u\|_{H^{1/2}} \leq C_K(\|Pu\|_0 + \|u\|_0).$$

(Apply Gårding's inequality (see Exercise 4.8) to $Q = P^*P - PP^*$.)

2) Show that the same inequality is true with H^s norms :

$$\|u\|_{H^{s+1/2}} \leq C_{K,s}(\|Pu\|_{H^s} + \|u\|_{H^s}).$$

3) Show that if $u \in \mathcal{D}'(X)$ and $Pu \in C^\infty(X)$ then $u \in C^\infty(X)$ (P is called hypoelliptic) :

a) Let $\tilde{X} \subset\subset X$ be open, so that $u \in H^{s_0}_{\text{loc}}(\tilde{X})$, for some $s_0 \in \mathbb{R}$. Let $\varphi \in C_0^\infty(\tilde{X})$. Show that $P\varphi u \in H^{s_0}(\mathbb{R}^n) \cap \mathcal{E}'(X)$.

b) Let $\chi \in C_0^\infty(B(0,1))$, $\int \chi\, dx = 1$, $\chi_\varepsilon(x) = \frac{1}{\varepsilon^n} \chi\left(\frac{x}{\varepsilon}\right)$.

Show that for ε small enough, $\chi_\varepsilon * \varphi u \in C_0^\infty(X)$ and $\text{supp}(\chi_\varepsilon * \varphi u)$ is in some fixed compact set. Use the inequality of 2) for $\chi_\varepsilon * \varphi u$ and show that

$$\|\chi_\varepsilon * \varphi u\|_{H^{s_0+1/2}} \leq C_1 \|P(\varphi u)\|_{H^{s_0}} + C_2 \|\varphi u\|_{H^{s_0}}.$$

Then show $\varphi u \in H^{s_0+1/2}$. (Hint : study the commutator of P and $\chi_\varepsilon *$, noting $\chi_\varepsilon * = \hat{\chi}(\varepsilon D)$.) Deduce that $u \in H^{s_0+1/2}_{\text{loc}}(\tilde{X})$.

c) By induction show that $u \in C^\infty(\tilde{X})$ and get $u \in C^\infty(X)$.

4) Apply the result of 3) to the operator Q in $\mathbb{R}^{n-1}$ of total symbol

$$\sigma_Q(x', \xi') = \sum_1^{n-1} \ell_j(x')\xi'_j + i\ell_n(x')|\xi'|$$

(see Exercise 4.3).

Find a simple geometric condition on the vector field $\ell = \sum_1^n \ell_j(x) D_{x_j}$ implying that Q is hypoelliptic.

Exercise 4.5

Let $A : C_0^\infty(\mathbb{R}^n) \to C^\infty(\mathbb{R}^n)$ with kernel $K(x,y)$. Assume

$$\sup_x \int |K(x,y)|\, dy < +\infty \quad \text{and} \quad \sup_y \int |K(x,y)|\, dx < +\infty.$$

Prove that A is continuous $L^2 \to L^2$ and

$$\|A\|_{\mathcal{L}(L^2,L^2)} \leq \left(\sup_x \int |K(x,y)|\, dy\right)^{1/2} \left(\sup_y \int |K(x,y)|\, dx\right)^{1/2}.$$

Exercise 4.6 (The Cotlar–Stein lemma)

1) Let $A_1, A_2, \ldots, A_N \in \mathcal{L}(E, F)$ where E, F are Hilbert spaces. Assume that

$$\sup_j \sum_{k=1}^N \|A_j^* A_k\|^{1/2} \leq M$$

$$\sup_j \sum_{k=1}^N \|A_j A_k^*\|^{1/2} \leq M$$

Let $A = \sum_{j=1}^{N} A_j$.

a) Show that $\|A\|^{2m} = \|(A^*A)^m\|$ and that
$$(A^*A)^m = \sum_{j_1, j_2, \ldots, j_{2m}} A_{j_1}^* A_{j_2} A_{j_3}^* \ldots A_{j_{2m-1}}^* A_{j_{2m}}.$$

b) Show that
$$\|A_{j_1}^* A_{j_2} \ldots A_{j_{2m}}\| \leq \|A_{j_1}^*\|^{\frac{1}{2}} \|A_{j_1}^* A_{j_2}\|^{\frac{1}{2}} \|A_{j_2} A_{j_3}^*\|^{\frac{1}{2}} \ldots \|A_{j_{2m-1}}^* A_{j_{2m}}\|^{\frac{1}{2}} \|A_{j_{2m}}\|^{\frac{1}{2}}.$$

c) Show that $\|A\|^{2m} \leq N M^{2m}$ for every $m \in \mathbb{N}$ and deduce that $\|A\| \leq M$.

2) Consider now an infinite sequence of operators $A_j \in \mathcal{L}(E, F)$, $j = 1, 2, \ldots$ such that
$$\sup_j \sum_{k=1}^{\infty} \|A_j^* A_k\|^{1/2} \leq M$$
$$\sup_j \sum_{k=1}^{\infty} \|A_j A_k^*\|^{1/2} \leq M$$

Show that $\sum_{j=1}^{\infty} A_j$ converges strongly and that the sum A satisfies $\|A\| \leq M$. (Consider the series $\sum A_j u$ first when $u \in \Sigma = \sum_k \operatorname{Im} A_k^*$, then when $u \in \overline{\Sigma}$, finally when $u \in \overline{\Sigma}^\perp$.)

The result of 2) is called the Cotlar–Stein lemma.

Exercise 4.7 (The theorem of Calderón and Vaillancourt)

Let $a(x, \xi) \in C^{\infty}(\mathbb{R}^n \times \mathbb{R}^n)$ satisfy
$$\forall \alpha, \beta \quad |\partial_x^\alpha \partial_\xi^\beta a(x, \xi)| \leq C_{\alpha\beta} \quad (*)$$

Consider the pseudodifferential operator $A \in L^0_{0,0}(\mathbb{R}^n)$ of the form
$$Au(x) = (2\pi)^{-n} \iint e^{i(x-y)\xi} a\left(\frac{x+y}{2}, \xi\right) u(y) \, dy \, d\xi.$$

(defined for $u \in \mathcal{S}(\mathbb{R}^n)$ as an iterated integral). We call a the Weyl symbol of A.

1) Let U_ν be the open ball of radius $R > 0$ and center $\nu \in \mathbb{Z}^{2n}$. Show that if R is large enough, there exists a function $\varphi_0 \in C_0^{\infty}(U_0)$ such that the functions $\varphi_\nu(x, \xi) = \varphi_0((x, \xi) - \nu) \in C_0^{\infty}(U_\nu)$, $\nu \in \mathbb{Z}^{2n}$, form a partition of unity : $1 = \sum_\nu \varphi_\nu(x, \xi)$.

Show that $a_\nu = a\varphi_\nu$ satisfy estimates like (*) with constants $\tilde{C}_{\alpha\beta}$, uniform in ν.

2) Let A_ν be the pseudodifferential operator with Weyl symbol a_ν. Show that $A_\nu A_\mu^*$ has kernel

$$(2\pi)^{-2n} \iiint e^{i(x\xi - y\eta)} e^{-iz(\xi - \eta)} a_\nu\left(\frac{x+z}{2}, \xi\right) \bar{a}_\mu\left(\frac{z+y}{2}, \eta\right) d\xi\, d\eta\, dz.$$

Use integrations by parts in ξ, η and z and use Exercise 4.5 to show that for every N

$$\|A_\nu A_\mu^*\|_{(L^2, L^2)} \leq C_N (1 + |\nu - \mu|)^{-N}.$$

3) Use the Cotlar–Stein lemma (Exercise 4.6) and show that A is continuous $L^2 \to L^2$.

Exercise 4.8 (The Gårding inequality)

Let $A \in L^{2m}_{cl}(X)$ be properly supported, $m \geq 0$. Assume that the principal symbol $a_{2m}(x, \xi)$ satisfies

$$\operatorname{Re} a_{2m} \geq \frac{1}{C_0} |\xi|^{2m}, \quad C_0 > 0.$$

Let $K \subset\subset X$ and $u \in C_0^\infty(X) \cap \mathcal{E}'(K)$.

1) Show that

$$\operatorname{Re}(Au \mid u) = (Bu \mid u)$$

where $B = \frac{1}{2}(A + A^*)$.

2) Show that there exists $C \in L^m_{cl}$ elliptic and $R \in L^{-\infty}$ such that

$$B = C^* C + R.$$

3) Show that there exists a constant $C(K)$ such that for every $u \in C_0^\infty(X) \cap \mathcal{E}'(K)$

$$\operatorname{Re}(Au \mid u) \geq \frac{1}{C(K)} \|u\|^2_{H^m} - C(K) \|u\|^2_{H^0}.$$

Exercise 4.9 (The sharp Gårding inequality)

Let $A \in L_{cl}^{2m}(X)$ be properly supported, $m \geq 0$ and assume that the principal symbol a_{2m} satisfies $\operatorname{Re} a_{2m} \geq 0$. The goal is to prove that for every compact set $K \subset\subset X$, $\exists C > 0$:

$$\operatorname{Re}(Au \mid u) \geq -C\|u\|_{H^{m-1/2}}^2 \qquad \forall u \in C_0^\infty(X) \cap \mathcal{E}'(K).$$

1) Let $\alpha = (\alpha_x, \alpha_\xi) \in X \times \mathbb{R}^n$.
Let $0 \leq \chi_{\alpha_x}(x) \in C_0^\infty(X)$ be equal to 1 near α_x and depend smoothly on α_x.
Let $f_\alpha(x) = C(n)^{\frac{1}{2}} |\alpha_\xi|^{\frac{n}{4}} e^{-\frac{1}{2}|\alpha_\xi|(x-\alpha_x)^2 + i(x-\alpha_x)\cdot\alpha_\xi}$ where $C(n) > 0$.
Let Π_α be the operator with kernel

$$\Pi_\alpha(x,y) = \chi_{\alpha_x}(x) f_\alpha(x) \overline{\chi_{\alpha_x}(y) f_\alpha(y)}.$$

Show that Π_α is properly supported in X and that $(\Pi_\alpha u \mid u) \geq 0$ for $u \in C_0^\infty(X)$.

2) Let $A \in L_{cl}^1(X)$ with symbol $a(x,\xi)$ and assume for simplicity that $a(x,\xi)$ vanishes for X outside some compact subset of X. Consider

$$\tilde{A} = \iint a(\alpha) \Pi_\alpha \, d\alpha$$

as an oscillatory integral.
Calculate for $\alpha_\xi \neq 0$

$$I(x, y, \alpha_\xi) = C(n) |\alpha_\xi|^{\frac{n}{2}} \int e^{-\frac{1}{2}|\alpha_\xi|[(x-\alpha_x)^2 + (y-\alpha_x)^2]} a(\alpha) \, d\alpha_x$$

and show that

$$I(x, y, \alpha_\xi) = \tilde{C}(n) b\left(\frac{x+y}{2}, \alpha_\xi\right) e^{-\frac{|\alpha_\xi|}{4}(x-y)^2}$$

where $b \in S_{cl}^1(\mathbb{R}^n \times \mathbb{R}^n)$ and $b_1 = a_1$.
Choose $C(n)$ correctly and show that $\tilde{A}u(x) = (2\pi)^{-n} \iint e^{i(x-y)\eta} c(x,\eta) u(y) \, d$
where $c(x,\eta) \sim \sum \frac{1}{\alpha!} D_y^\alpha \partial_\eta^\alpha \left(e^{-\frac{1}{4}|\eta|(x-y)^2} b\left(\frac{x+y}{2}, \eta\right) \right)\big|_{y=x}$ in $S_{1,\frac{1}{2}}^1(X \times \mathbb{R}^n)$.
Show that $c(x,\eta) \in S_{cl}^1$ with $c_1 = a_1$ and finally that $\tilde{A} - A \in L_{cl}^0$. Extend the result to the case when the support condition on a is dropped.

3) Let now $A \in L_{cl}^1$ with $\operatorname{Re} a_1 \geq 0$. Write $B = \frac{1}{2}(A + A^*)$ and let

$$\tilde{B} = \iint \operatorname{Re} a_1(\alpha) \Pi_\alpha \, d\alpha.$$

Show that $\tilde{B} - B \in L^0_{cl}$ and $(\tilde{B}u \mid u) \geq 0$.
Show then that

$$\operatorname{Re}(Au \mid u) \geq -C(K)\|u\|_0^2, \quad u \in C_0^\infty(X) \cap \mathcal{E}'(K), \quad K \subset\subset X.$$

4) Let now $A \in L^{2m}_{cl}$, $m \geq 0$ with $\operatorname{Re} a_{2m} \geq 0$. Let $\Lambda_{m-\frac{1}{2}}$ be a positive elliptic properly supported operator of order $m - \frac{1}{2}$. Write

$$A \equiv \Lambda_{m-\frac{1}{2}} B \Lambda_{m-\frac{1}{2}} \bmod L^{-\infty}.$$

Show that $\operatorname{Re}(Au \mid u) \geq \operatorname{Re}(B\Lambda_{m-\frac{1}{2}}u \mid \Lambda_{m-\frac{1}{2}}u) - C\|u\|^2_{H^{m-\frac{1}{2}}}$.
Get that $\operatorname{Re}(Au \mid u) \geq -C\|u\|^2_{H^{m-\frac{1}{2}}}$.

Exercise 4.10 (Elliptic pseudodifferential operators on a compact manifold)

Let M be a compact smooth manifold and let $P \in L^m_{cl}(M)$ with principal symbol $p \in C^\infty(T^*M\backslash 0)$ positively homogeneous of degree m.
P is *elliptic* if $p(x,\xi) \neq 0 \quad \forall x \in M, \quad \forall (x,\xi) \in T^*M\backslash 0$.

1) Let $P \in L^m_{cl}$ be elliptic. Find $Q \in L^{-m}_{cl}$ such that

$$PQ = I - R_1$$
$$QP = I - R_2$$

with $R_j \in L^{-\infty}(M)$, $j = 1, 2$.

2) Let $R \in L^{-\infty}(M)$. Show that $R : H^{s_1} \to H^{s_2}$ is compact for all $s_1, s_2 \in \mathbb{R}$.

3) Consider $P = P_s : H^{s+m} \to H^s$.
Show that $\operatorname{Ker} P_s$ is of finite dimension and that $\operatorname{Ker} P_s \subset C^\infty(M)$.

4) Let L be a closed subspace of $H^{s+m}(M)$ such that $H^{s+m}(M) = L \oplus \operatorname{Ker} P_s$.

 a) Show that $\|u\|_{H^{s+m}} \leq C \|Pu\|_{H^s}$ for every $u \in L$. (Assume there exists a sequence $u_j \in L$, $\|u_j\|_{H^{s+m}} = 1$ with $Pu_j \to 0$ in H^s and find a contradiction.)

 b) Show that $\operatorname{Im} P_s$ is a closed subspace. (Consider $v_j \in \operatorname{Im} P$ with $v_j \to v_0$ in H^s and $u_j \in L$ with $Pu_j = v_j$.)

 c) Show that $(\operatorname{Im} P_s)^\perp = \operatorname{Ker} P_s^*$ with $P_s^* : H^{-s} \to H^{-s-m}$ being the adjoint of P_s. Show that $\operatorname{Im} P_s = \{u \in H^s; (u \mid v_j) = 0, j = 1, \ldots, N\}$ where $\{v_1, \ldots, v_N\}$ is a basis of $\operatorname{Ker} P_s^*$. Show that one can define the index of P as

$$\operatorname{ind} P = \operatorname{ind} P_s = \dim \operatorname{Ker} P_s - \operatorname{codim} \operatorname{Im} P_s$$

Notes

This chapter contains some standard results for pseudodifferential operators. The proof of Theorem 4.5 (cf. Exercise 4.8) is taken from [Hö2] and is an elaboration of a corresponding one in [KN]. For the Cotlar–Stein lemma and the Calderón–Vaillancourt theorem, see [Cot], [KSt], [CaV1,2], [Bo].

For the Gårding inequality and its various extensions, see [Gå], [Hö6], [LN], [CF], [T], [FP1,2], [Hö4]. Exercise 4.9 follows an idea of [CF] ; see also [T].

5 Local symplectic geometry I (Hamilton–Jacobi theory)

We will assume the reader is familiar with some basic notions of differential geometry such as : manifold, tangent and cotangent vectors, differential form, (vector) bundle. Nevertheless we recall briefly (and non-systematically) some of these notions.

Tangent and cotangent vectors

Let X be a smooth manifold of dimension n. Let $x_0 \in X$. If $\gamma, \tilde{\gamma} :]-1,1[\to X$ are two C^1-curves with $\gamma(0) = x_0$, $\tilde{\gamma}(0) = x_0$, we say that $\gamma, \tilde{\gamma}$ are equivalent if $\|\gamma(t) - \tilde{\gamma}(t)\| = o(t)$, $t \to 0$. (Here we choose some local coordinates $x_1, \ldots, x_n$ near x_0 so that $\|\gamma(t) - \tilde{\gamma}(t)\|$ is well defined. The choice of these coordinates does not affect the definition.) A tangent vector is by definition an equivalence class. If γ is a curve as above we denote by $\dot{\gamma}(0)$ or $\left(\frac{d}{dt}\right)_{t=0} \gamma(t)$ the corresponding tangent vector. The set of tangent vectors at a point $x_0 \in X$ is denoted by $T_{x_0} X$.

If $f, \tilde{f} : X \to \mathbb{R}$ are two C^1-functions, we say that f and $\tilde{f}$ are equivalent if $f(x) - f(x_0) - (\tilde{f}(x) - \tilde{f}(x_0)) = o(\|x - x_0\|)$, $x \to x_0$. We let $df(x_0)$ denote the equivalence class of f. It is called a differential 1-form at x_0 or a cotangent vector at x_0 ; also it is the differential of f at x_0.

$T^*_{x_0} X$ and $T_{x_0} X$ are n-dimensional (real) vector spaces dual to each other. The duality is given by

$$\langle df(x_0), \dot{\gamma}(0) \rangle = \left(\frac{d}{dt}\right)_{t=0} f(\gamma(t)).$$

If $x_1, \ldots, x_n$ are local coordinates, defined in a neighborhood of x_0, then $dx_1(x_0), \ldots, dx_n(x_0)$ (or $dx_1, \ldots, dx_n$ for short) is a basis in $T^*_{x_0} X$ and the corresponding dual basis is (by definition) $\frac{\partial}{\partial x_1}(x_0), \ldots, \frac{\partial}{\partial x_n}(x_0)$ (or $\frac{\partial}{\partial x_1}, \ldots, \frac{\partial}{\partial x_n}$ for short).

The sets $TX = \bigcup_{x_0 \in X} T_{x_0} X$ and $T^*X = \bigcup_{x_0 \in X} T^*_{x_0} X$ are vector bundles and in particular C^∞-manifolds. If $x_1, \ldots, x_n$ are local coordinates on X then we get corresponding local coordinates $(x, t) = (x_1, \ldots, x_n, t_1, \ldots, t_n)$ on TX and $(x, \xi) = (x_1, \ldots, x_n, \xi_1, \ldots, \xi_n)$ on T^*X by representing $\nu \in TX$ and $\rho \in T^*X$ by their base point x (described by $(x_1, \ldots, x_n)$) so that $\nu \in T_x X$, $\rho \in T^*_x X$ and then writing $\nu = \sum t_j \frac{\partial}{\partial x_j}(x)$, $\rho = \sum \xi_j dx_j$. The local coordinates $(x_1, \ldots, x_n, \xi_1, \ldots, \xi_n)$ are called canonical (local) coordinates on T^*X. If $y_1, \ldots, y_n$ is a second system of local coordinates, then in the intersection of the two open sets in X parametrized by the two systems of coordinates we have the relations $t = \dfrac{\partial x}{\partial y} s$, $\eta = {}^t\!\left(\dfrac{\partial x}{\partial y}\right) \xi$ for the corresponding local coordinates (x,t), (y,s) on TX and (x,ξ), (y,η) on T^*X. Here $\dfrac{\partial x}{\partial y} = \left(\dfrac{\partial x_j}{\partial y_k}\right)$ is the standard Jacobian matrix.

If $\rho \in T^*X$ we let $\pi(\rho) \in X$ be the corresponding base point. A section in T^*X is a map $\omega : X \to T^*X$ with $\pi \circ \omega(x) = x$, $\forall x \in X$. (The same definition can be given for TX or for any vector bundle.) Sections in T^*X are called differential (1-)forms, and sections in TX are called vector fields. If nothing else is specified they will be assumed to be of class C^∞. Also most of the time we only consider sections which are defined locally.

A vector field can be written in local coordinates as $\nu = \sum\limits_{1}^{n} a_j(x)\dfrac{\partial}{\partial x_j}$ and a differential 1-form as $\omega = \sum\limits_{1}^{n} \xi_j(x)\, dx_j$.

If Y is a second manifold and $f : Y \to X$ is a map of class C^1, $y_0 \in Y$, $x_0 = f(y_0) \in X$, then we have a natural map $f_* = df : T_{y_0}Y \to T_{x_0}X$ which in local coordinates is given by the ordinary Jacobian matrix. The adjoint is $f^* : T^*_{x_0}X \to T^*_{y_0}Y$ and we notice that $d(u \circ f)(y_0) = f^*(du(x_0))$ if u is a C^1 function on X. If Z is a third manifold, $g : Z \to Y$ in C^1 and $z_0 \in Z$, $g(z_0) = y_0$, then $(f \circ g)_* = f_* \circ g_*$, $(f \circ g)^* = g^* \circ f^*$. When passing to sections we see that if ω is a 1-form on X then $f^*\omega$ is a well defined 1-form on Y (this is the pull-back of ω by means of f). Notice that the corresponding push-forward $f_*\nu$ of a vector field ν on Y can be defined if f is a C^1 diffeomorphism but not in general.

If $\gamma :]a, b[\to X$ is a C^1-curve and $t_0 \in]a, b[$ we define its tangent at $\gamma(t_0)$ as $\gamma_*\!\left(\dfrac{\partial}{\partial t}\right) \stackrel{\text{def}}{=} \dfrac{\partial \gamma}{\partial t}(t_0) = \dot\gamma(t_0)$. (This definition coincides with the earlier one.) If ν is a C^∞ vector field on X then for every $x_0 \in X$ we can find $T_+(x_0)$, $T_-(x_0) \in]0, +\infty]$ such that we have a unique smooth curve

$$]-T_-(x_0), T_+(x_0)[\ni t \mapsto \gamma(t) = \exp t\nu(x_0) \in X$$

with $\gamma(0) = x_0$, $\dot\gamma(t) = \nu(\gamma(t))$.

Choosing $T_+(x_0)$, $T_-(x_0)$ maximal, we get a smooth map

$$\Phi : \{(t, x) \in \mathbb{R} \times X \,;\, -T_-(x) < t < T_+(x)\} \to X, \quad \Phi(t, x) = \exp(t\nu)(x),$$

where $\Phi(0, x) = x$, $\dfrac{\partial \Phi(t, x)}{\partial t} = \nu(\Phi(t, x))$ and $T_+(x), T_-(x)$ are lower semi-continuous. We have

$$\exp t\nu \circ \exp s\nu(x) = \exp(t + s)\nu(x),$$

for t, s such that both members are defined.

The canonical 1- and 2-forms

Let $\pi : T^*X \to X$ be the natural projection $((x,\xi) \mapsto x$ in canonical coordinates). For $\rho \in T^*X$ we consider $\pi^* : T^*_{\pi(\rho)}X \to T^*_\rho(T^*X)$ and since $\rho \in T^*_{\pi(\rho)}X$ we can define the canonical 1-form $\omega_\rho \in T^*_\rho(T^*X)$ by $\omega_\rho = \pi^*(\rho)$. Varying ρ we get a smooth 1-form on T^*X. In canonical coordinates we get
$$\omega = \sum_1^n \xi_j \, dx_j.$$

We next recall a few facts about forms of higher degree. If L is a finite-dimensional real vector space and L^* is the dual space, then we have a natural duality between the k-fold exterior product spaces $\Lambda^k L$ and $\Lambda^k L^*$, given by

$$\langle u_1 \wedge \ldots \wedge u_k, \, v_1 \wedge \ldots \wedge v_k \rangle = \det(\langle u_j, v_k \rangle), \quad u_j \in L, \, v_k \in L^*.$$

If M is a C^∞ manifold of dimension m then a differential k-form is a section v of the vector bundle $\Lambda^k T^*M$. In local coordinates $x_1, \ldots, x_m$:

(5.1) $$v = \sum_{|I|=k} v_I(x) \, dx^I$$

where in general $I = (i_1, \ldots, i_\ell) \in \{1, \ldots, m\}^\ell$, $|I| = \ell$, $dx^I = dx_{i_1} \wedge \ldots \wedge dx_{i_\ell}$.

(The representation (5.1) becomes unique if we restrict the I's to those with $i_1 < i_2 < \ldots < i_k$.) If v is a k-form of class C^1 locally given by (5.1), we define the $(k+1)$-form

(5.2) $$dv = \sum_{|I|=k} dv_I \wedge dx^I \quad \text{(the exterior differential of } v\text{)}.$$

This definition does not depend on the choice of local coordinates or on how we choose the representation (5.1). We have the following facts

(5.3) $$d^2 = 0.$$

(5.4) If w is a C^∞ $(k+1)$-form which is closed in the sense that $dw = 0$, then in every open set in M diffeomorphic to a ball, we can find a smooth k-form v such that $dv = w$.

(5.5) If $f : Y \to X$ is a smooth map between two smooth manifolds then there is a unique way of extending the pull-back f^* of 1-forms to k-forms by multilinearity. If v is a smooth k-form on X, then $df^*v = f^*dv$.

We now return to the canonical 1-form ω on T^*X and define the canonical 2-form σ on T^*X as $\sigma = d\omega$. In canonical coordinates :

(5.6) $$\sigma = \sum_1^n d\xi_j \wedge dx_j.$$

For $\rho \in T^*X$, $\sigma_\rho \in \Lambda^2 T^*_\rho(T^*X)$ can be viewed as a linear form on $\Lambda^2 T_\rho(T^*X)$ or equivalently as an antisymmetric bilinear form on $T_\rho(T^*X) \times T_\rho(T^*X)$ given by

$$\sigma_\rho(t,s) = \langle \sigma_\rho, t \wedge s \rangle, \quad t,s \in T_\rho(T^*X).$$

In canonical coordinates we write $t = (t_x, t_\xi)$ $\left(t = \sum t_{x_j} \frac{\partial}{\partial x_j} + \sum t_{\xi_j} \frac{\partial}{\partial \xi_j}\right)$, $s = (s_x, s_\xi)$ and we get

$$\sigma_\rho(t,s) = \langle t_\xi, s_x \rangle - \langle s_\xi, t_x \rangle = \sum t_{\xi_j} s_{x_j} - s_{\xi_j} t_{x_j}.$$

From this it is clear that σ_ρ is a non-degenerate bilinear form and we therefore have a bijection $H : T^*_\rho(T^*X) \to T_\rho(T^*X)$ defined by

$$\sigma(s, Hu) = \langle s, u \rangle, \quad s \in T_\rho(T^*X), \quad u \in T^*_\rho(T^*X).$$

In canonical coordinates, if $u = u_x\, dx + u_\xi\, d\xi$ $\left(= \sum u_{x_j}\, dx_j + u_{\xi_j}\, d\xi_j\right)$ we get $Hu = u_\xi \frac{\partial}{\partial x} - u_x \frac{\partial}{\partial \xi}$.

If $f(x, \xi)$ is of class C^1 on (some open set in) T^*X, we define its Hamilton (vector) field by $H_f = H(df)$. In canonical coordinates,

$$H_f = \sum_1^n \frac{\partial f}{\partial \xi_j} \frac{\partial}{\partial x_j} - \frac{\partial f}{\partial x_j} \frac{\partial}{\partial \xi_j}.$$

In a more sophisticated way, let M be a manifold, $\rho \in M$, $t \in T_\rho M$ and define $t \lrcorner : \Lambda^k T^*_\rho M \to \Lambda^{k-1} T^*_\rho M$ as the adjoint of the left exterior multiplication $t \wedge : \Lambda^{k-1} T_\rho M \to \Lambda^k T_\rho M$. Then with $M = T^*X$, the Hamilton field is defined by the pointwise relation

(5.7) $$H_f \lrcorner \sigma = -df.$$

If f, g are two C^1 functions defined on the same open set in T^*X, we define their Poisson bracket as the continuous function

$$\{f, g\} = H_f(g) = \langle H_f, dg \rangle = \sigma(H_f, H_g),$$

where in the second expression we view H_f as a first-order differential operator. In canonical coordinates,

$$\{f, g\} = \sum \frac{\partial f}{\partial \xi_j} \frac{\partial g}{\partial x_j} - \frac{\partial f}{\partial x_j} \frac{\partial g}{\partial \xi_j}.$$

Notice that $\{f, g\} = -\{g, f\}$ and in particular $\{f, f\} = 0$.

Example 5.1 If $X \subset \mathbb{R}^n$ is open, we can identify T^*X with $X \times \mathbb{R}^n$ (by means of the canonical coordinates induced by the standard coordinates on X), and if $P \in L_{cl}^{m'}(X)$, $Q \in L_{cl}^{m''}(X)$ are properly supported, then $R \stackrel{\text{def}}{=} [P,Q] \in L_{cl}^{m'+m''-1}(X)$ and for the corresponding homogeneous principal symbols p, q, r we have $r = \dfrac{1}{i}\{p,q\}$.

Lie derivatives

Let v be a C^∞ vector field on a manifold M and let ω be a C^∞ k-form on M. Then the Lie derivative of ω along v is defined pointwise by

$$\mathcal{L}_v \omega = \left(\frac{d}{dt}\right)_{t=0} ((\exp tv)^* \omega).$$

If u is a second smooth vector field on M we also define

$$\mathcal{L}_v u = \left(\frac{d}{dt}\right)_{t=0} ((\exp -tv)_* u).$$

In the last definition we need to observe that the push-forward of a vector field by means of a local diffeomorphism can be defined locally. We have the following identities :

1) When ω is a 0-form and hence a function, then $\mathcal{L}_v \omega = v(\omega)$.

2) $\mathcal{L}_v u = [v, u] = vu - uv$, where u, v are viewed as first-order differential operators in the last two expressions.

3) $\mathcal{L}_v (d\omega) = d(\mathcal{L}_v \omega)$

4) $\mathcal{L}_v(\omega_1 \wedge \omega_2) = (\mathcal{L}_v \omega_1) \wedge \omega_2 + \omega_1 \wedge (\mathcal{L} \omega_2)$

5) $\mathcal{L}_v(u \lrcorner \omega) = (\mathcal{L}_v u) \lrcorner \omega + u \lrcorner (\mathcal{L}_v \omega)$

6) $\mathcal{L}_v \omega = v \lrcorner d\omega + d(v \lrcorner \omega)$

7) $\mathcal{L}_{v_1+v_2} = \mathcal{L}_{v_1} + \mathcal{L}_{v_2}$.

Lemma 5.2 *If f is a C^∞ function on some open set in T^*X, then $\mathcal{L}_{H_f} \sigma = 0$.*

Proof: $\mathcal{L}_{H_f} \sigma = H_f \lrcorner d\sigma + d(H_f \lrcorner \sigma) = H_f \lrcorner d^2\omega - d^2 f = 0$. $\square$

Locally we can define the maps $\Phi_t = \exp t H_f$ when $|t|$ is sufficiently small. We have then $\Phi_t^* \sigma = \sigma$. In fact, we have pointwise

$$\frac{d}{dt} \Phi_t^* \sigma = \left(\frac{d}{ds}\right)_{s=0} \Phi_t^* \Phi_s^* \sigma = \Phi_t^* \mathcal{L}_{H_f} \sigma = 0.$$

Lagrangian manifolds

A submanifold $\Lambda \subset T^*X$ is called a Lagrangian manifold if $\dim \Lambda = \dim X$ and $\sigma|_\Lambda = 0$. In general we define the restriction of a differential k-form to a submanifold as the pull-back of this form by means of the natural inclusion map. Viewing $(\sigma|_\Lambda)_\rho$, $\rho \in \Lambda$ as a bilinear form on $T_\rho\Lambda \times T_\rho\Lambda$ we simply have $(\sigma|_\Lambda)_\rho(t,s) = \sigma_\rho(t,s)$, $t,s \in T_\rho\Lambda$, where $T_\rho\Lambda$ is identified with a subspace of $T_\rho T^*X$, (namely the image of $T_\rho\Lambda$ by the differential of the natural inclusion map). If $T_\rho\Lambda^\perp$ denotes the orthogonal space of $T_\rho\Lambda$ in $T_\rho T^*X$ with respect to the bilinear form σ_ρ, then we see that a submanifold $\Lambda \subset T^*X$ is Lagrangian if and only if $T_\rho\Lambda^\perp = T_\rho\Lambda$ for every $\rho \in \Lambda$.

Theorem 5.3 *Let $\Lambda \subset T^*X$ be a submanifold with $\dim \Lambda = \dim X$ and such that $\pi|_\Lambda : \Lambda \to X$ is a local diffeomorphism (in the sense that every point ρ in Λ has a neighborhood in Λ which is mapped diffeomorphically by $\pi|_\Lambda$ onto a neighborhood of $\pi(\rho)$). Then Λ is Lagrangian if and only if for each point ρ of Λ we can find a (real) C^∞ function $\varphi(x)$ defined near the projection of ρ, such that Λ coincides near ρ with the manifold $\{(x, d\varphi(x)) \,;\, x \in$ some neighborhood of $\pi(\rho)\}$.*

Proof: If ω is the canonical 1-form, we notice that $d(\omega|_\Lambda) = \sigma|_\Lambda$. Therefore the following three statements are equivalent.

1) Λ is Lagrangian.

2) $\omega|_\Lambda$ is closed (i.e. $d(\omega|_\Lambda) = 0$).

3) Locally on Λ we can find a smooth function φ with $\omega|_\Lambda = d\varphi$.

If $x_1, \ldots, x_n$ are local coordinates on X, we can also view them (or rather their compositions with π) as local coordinates on Λ, and represent Λ by $\xi = \xi(x)$ in the corresponding canonical coordinates. Then 3) is equivalent to $\xi_j(x) = \dfrac{\partial \varphi(x)}{\partial x_j}$ i.e. $\sum \xi_j(x)\, dx_j = d\varphi$. □

Hamilton–Jacobi equations

These equations are of the form $p(x, \varphi'_x) = 0$, where p is a real-valued C^∞ function defined on some open subset of T^*X. Here we shall also assume that $dp(x,\xi) \neq 0$ when $p = 0$. The basic idea here is to consider the Lagrangian manifold Λ associated with φ, and to try to construct such a manifold inside the hypersurface H defined by $p(x,\xi) = 0 : \Lambda \subset H$. If $\rho \in \Lambda$, we shall then have $T_\rho\Lambda \subset T_\rho H$ (considering these tangent spaces as subspaces of $T_\rho T^*X$), and hence $T_\rho H^\perp \subset T_\rho\Lambda$ (since $T_\rho\Lambda^\perp = T_\rho\Lambda$). Now $T_\rho H^\perp = \mathbb{R}\, H_p$ so we must have $H_p \in T_\rho\Lambda$ at every point $\rho \in \Lambda$, or in other words, that H_p must be tangent to Λ at every point of Λ.

Proposition 5.4 Let $\Lambda' \subset H$ be an isotropic submanifold (in the sense that $\sigma|_{\Lambda'} = 0$) of dimension $n-1$ passing through some given point $\rho_0 \in H$ and such that $H_p(\rho_0) \notin T_{\rho_0}\Lambda'$. Then in a neighborhood of ρ_0, we can find a Lagrangian manifold Λ such that $\Lambda' \subset \Lambda \subset H$ (in that neighborhood).

Proof: According to the observation above it is natural to try

$$\Lambda = \{\exp(tH_p)(\rho)\,;\, |t| < \varepsilon,\, \rho \in \Lambda',\, |\rho - \rho_0| < \varepsilon\}$$

for some sufficiently small $\varepsilon > 0$. (Here $|\rho - \rho_0|$ is well defined if we choose some local canonical coordinates.) Then $\Lambda' \subset \Lambda$ (near ρ_0) and since H_p is tangent to H (by the relation $H_p p = 0$) we also have $\Lambda \subset H$. From the assumption $H_p(\rho_0) \notin T_{\rho_0}\Lambda'$ and the implicit function theorem, it also follows that Λ is a smooth manifold of dimension n.

In order to verify that Λ is Lagrangian, we first take $\rho \in \Lambda'$ ($|\rho - \rho_0| < \varepsilon$) and consider $T_\rho \Lambda = T_\rho \Lambda' \oplus \mathbb{R} H_p$. Then $\sigma_\rho|_{T_\rho \Lambda \times T_\rho \Lambda} = 0$ since $\sigma_\rho|_{T_\rho \Lambda' \times T_\rho \Lambda'} = 0$, $\sigma_\rho(H_p, H_p) = 0$, $\sigma_\rho(t, H_p) = 0$, $t \in T_\rho \Lambda'$. (The last fact follows from $\sigma_\rho(t, H_p) = \langle t, dp\rangle = 0$, for all $t \in T_\rho H$.)

More generally, at the point $\rho_t = \exp t H_p(\rho)$, $\rho \in \Lambda'$, we have $T_{\rho_t}(\Lambda) = \exp(tH_p)_*(T_\rho \Lambda)$ and for $u, v \in T_\rho \Lambda$ we get (using that $\exp(tH_p)^* \sigma_{\rho_t} = \sigma_\rho$) :

$$\sigma_{\rho_t}(\exp(tH_p)_* u,\, \exp(tH_p)_* v) = \sigma_\rho(u,v) = 0.$$

$\square$

In the following we write $x = (x', x_n) \in \mathbb{R}^n$, $x' = (x_1, \ldots, x_{n-1}) \in \mathbb{R}^{n-1}$.

Theorem 5.5 Let $p(x, \xi)$ be a real-valued C^∞ function, defined in a neighborhood of some point $(0, \xi_0) \in T^*\mathbb{R}^n$, such that $p(0, \xi_0) = 0$, $\dfrac{\partial p}{\partial \xi_n}(0, \xi_0) \neq 0$. Let $\psi(x')$ be a real-valued C^∞ function defined near 0 in $\mathbb{R}^{n-1}$ such that $\dfrac{\partial \psi}{\partial x'}(0) = \xi_0'$. Then there exists a real-valued smooth function $\varphi(x)$, defined in a neighborhood of $0 \in \mathbb{R}^n$, such that in that neighborhood :

(5.8) $\quad p(x, \varphi_x'(x)) = 0,\quad \varphi(x', 0) = \psi(x'),\quad \varphi_x'(0) = \xi_0.$

If $\tilde\varphi(x)$ is a second function with the same properties, then $\varphi(x) = \tilde\varphi(x)$ in some neighborhood of 0.

Proof: In a suitable neighborhood of $(0, \xi_0) \in \mathbb{R}^{n-1} \times \mathbb{R}^n$ we have $p(x', 0, \xi) = 0$ if and only if $\xi_n = \lambda(x', \xi')$, where λ is a real-valued C^∞ function, with $\lambda(0, \xi_0') = (\xi_0)_n$. Let

$$\Lambda' = \left\{(x,\xi)\,;\, x_n = 0,\ \xi' = \frac{\partial \psi}{\partial x'}(x'),\ \xi_n = \lambda(x', \xi'),\ x' \in \text{neigh}(0)\right\}$$

(where "$x' \in \text{neigh}(0)$" means that x' belongs to some sufficiently small neighborhood of 0).

Then $\Lambda' \subset p^{-1}(0)$ is isotropic of dimension $n-1$ and H_p is nowhere tangent to Λ' since H_p has a component $\dfrac{\partial p}{\partial \xi_n}\dfrac{\partial}{\partial x_n}$ with $\dfrac{\partial p}{\partial \xi_n} \neq 0$. Let $\Lambda \subset p^{-1}(0)$ be a Lagrangian manifold as in Proposition 5.4. The differential of $\pi|_{\Lambda}: \Lambda \to \mathbb{R}^n$ is bijective at $(0,\xi_0)$ so if we restrict our attention to a sufficiently small neighborhood of $(0,\xi_0)$, Λ becomes of the form $\xi = \varphi'(x)$, $x \in \text{neigh}(0)$. Hence $p(x,\varphi'(x)) = 0$, $\varphi'(0) = \xi_0$. Since $\Lambda' \subset \Lambda$ we get $\dfrac{\partial \psi}{\partial x'}(x') = \dfrac{\partial \varphi}{\partial x'}(x',0)$, and modifying φ by a constant we get $\varphi(x',0) = \psi(x')$. The uniqueness property is left to the reader as an exercise. $\square$

We can view Λ as a union of integral curves of H_p passing through Λ'. The projection of such an integral curve is an integral curve of the field
$$\nu = \sum_1^n \frac{\partial p}{\partial \xi_j}(x,\varphi'_x)\frac{\partial}{\partial x_j}$$
(which via $\pi|_\Lambda$ can be identified with $H_p|_\Lambda$). If $q(x,\xi) = \sum_1^n \dfrac{\partial p}{\partial \xi_j}(x,\xi)\xi_j$ we get the highly trivial identity

$$\left(\sum_1^n \frac{\partial p}{\partial \xi_j}(x,\varphi'_x)\frac{\partial}{\partial x_j}\right)\varphi = q(x,\varphi'_x).$$

Hence if $x = x(t)$ is an integral curve of ν with $x_n(0) = 0$, then we get $\varphi(x(t)) = \psi(x'(0)) + \int_0^t q(x(s),\xi(s))\,ds$, where $\xi(s) = \varphi'(x(s))$, so that $s \mapsto (x(s),\xi(s))$ is the integral curve of H_p with $x_n(0) = 0$, $\xi'(0) = \dfrac{\partial \psi}{\partial x'}(x'(0))$, $\xi_n(0) = \lambda(x'(0),\xi'(0))$. In particular, if p is positively homogeneous of degree m, then by the Euler homogeneity relations, $q(x,\xi) = mp(x,\xi) = 0$ on Λ and we obtain $\varphi(x(t)) = \psi(x'(0))$.

If $\psi = \psi_\alpha$ depends smoothly on some parameters $\alpha \in \mathbb{R}^k$, then $\varphi = \varphi(x,\alpha)$ will be a smooth function of (x,α), and differentiating the equation $p(x,\varphi'_x) = 0$ with respect to α we get

$$\sum p^{(j)}(x,\varphi'_x)\frac{\partial}{\partial x_j}\frac{\partial \varphi}{\partial \alpha} = 0$$

so that $\dfrac{\partial \varphi}{\partial \alpha}$ is constant along the bicharacteristic curves (without any homogeneity assumption). Here we use standard notation and terminology from partial differential equations: $p^{(j)}(x,\xi) = \dfrac{\partial p}{\partial \xi_j}(x,\xi)$, $p_{(j)}(x,\xi) = \dfrac{\partial p}{\partial x_j}(x,\xi)$, a bicharacteristic curve is the x-space projection of a bicharacteristic strip, the latter is by definition an integral curve of H_p along which p vanishes.

Exercises

Exercise 5.1

Let v be a vector field defined in a neighborhood of $x_0 \in \mathbb{R}^n$ and satisfying $v(x_0) \neq 0$.

Show that there exist local coordinates $x_1, \ldots, x_n$ in a neighborhood of x_0, such that $v = \dfrac{\partial}{\partial x_n}$.

Exercise 5.2

Let $v = \sum_1^n a_j(x) \dfrac{\partial}{\partial x_j}$ be a C^∞ vector field, defined in a neighborhood of $0 \in \mathbb{R}^n$, and suppose $v(0) = 0$. We introduce the linearized vector field of v at 0 as $\sum\sum \dfrac{\partial a_j}{\partial x_k}(0) x_k \dfrac{\partial}{\partial x_j}$, and we consider $A = \left(\dfrac{\partial a_j}{\partial x_k}(0)\right)$ as an element of $\text{End}(T_0 \mathbb{R}^n)$ and let $\sigma(A) \subset \mathbb{C}$ be the set of eigenvalues of A.

a) If $\lambda \in \sigma(A) \Longrightarrow \text{Re}\lambda < 0$, show that there exists $C > 0$ such that if $|x| \leq \dfrac{1}{C}$ and $t \geq 0$ then $|\exp(tv)(x)| \leq C e^{-t/C} |x|$.

b) Suppose that $n = 2$ and that A has one strictly positive eigenvalue and one strictly negative eigenvalue. Study the shape of the integral curves near $(0,0)$.

c) Also with $n = 2$, study the integral curves of $x_1 \partial_{x_2} - x_2 \partial_{x_1} + w$ in the following cases.

    α) $w = 0$
    β) $w = -|x|^2(x_1 \partial_{x_1} + x_2 \partial_{x_2})$
    γ) $w = |x|^2(x_1 \partial_{x_1} + x_2 \partial_{x_2})$.

Exercise 5.3

Let σ be the canonical 2-form on T^*X, and ν a vector field defined on some open subset of T^*X, diffeomorphic to a ball. Show that ν is a Hamilton field if and only if $\mathcal{L}_\nu \sigma = 0$.

Exercise 5.4

A smooth submanifold Σ of T^*X is called *involutive* if it has the property:

i) $\{u, v\} = 0$ on Σ for all smooth functions u, v on T^*X which vanish on Σ.

1) Show that this definition is equivalent to

ii) $\quad T_\rho \Sigma^\perp \subset T_\rho \Sigma \quad \forall \rho \in \Sigma$
(where $\perp$ indicates the orthogonal with respect to the canonical 2-form σ)

and also equivalent to

iii) $\quad u = 0$ on $\Sigma \Longrightarrow H_u$ is tangent to Σ
(where H_u is the Hamiltonian vector field of u).

2) Show that if Σ is involutive then $\dim \Sigma \geq \dim X$.

3) Show that Σ is Lagrangian if and only if Σ is involutive and of dimension $\dim X$.

Exercise 5.5

Let $P \in L^m_{\text{cl}}(M)$ where M is a C^∞ compact manifold as in Exercise 3.4. Let $p_j \in C^\infty(X_j \times \dot{\mathbb{R}}^n)$ be the principal symbol of the representative P_j of P.

1) Show that there exists $p \in C^\infty(T^*M \backslash 0)$ positively homogeneous of degree m such that

$$p_j(\kappa_j(x), \xi) = p(x, {}^t\kappa'_j(x)(\xi)), \quad x \in M_j, \quad \xi \in \mathbb{R}^n.$$

By definition p is the *principal symbol* of P.

2) Show that the principal symbol of P can be defined by

$$\lim_{\lambda \to +\infty} \lambda^{-m} e^{-i\lambda\varphi(x)} P(e^{i\lambda\varphi(x)} u) = u(x) p(x, \varphi'(x))$$

if $u \in C^\infty(M)$, $\varphi \in C^\infty(M, \mathbb{R})$, $\varphi'(x) \neq 0$ on $\operatorname{supp} u$. (We may choose local coordinates so that $\varphi(x)$ is affine linear.)

3) What is the principal symbol of $P \circ Q$ and $[P, Q]$ if $P \in L^m_{\text{cl}}(M)$, $Q \in L^m_{\text{cl}}(M)$?

4) Extend these results to operators in $L^m_\rho(M)$, $\rho > \dfrac{1}{2}$.

Exercise 5.6

Let $P \in L^m_{\text{cl}}(\mathbb{R}^n)$ with symbol $p(x, \xi) = p_m(x, \xi) + p_{m-1}(x, \xi) + \ldots + p_{m-j}(x, \xi) + \ldots$, p_{m-j} positively homogeneous of degree $m-j$ in ξ.

Let us denote $\sigma(P) = p_m$, $\operatorname{sub} P = p_{m-1} - \dfrac{1}{2i} \sum_{j=1}^n \dfrac{\partial^2 p_m}{\partial \xi_j \, \partial x_j}$.

1) Show that if $P \in L_{cl}^m$, $Q \in L_{cl}^{m'}$:

$$\mathrm{sub}\,(P \circ Q) = \sigma(P)\mathrm{sub}\,Q + \sigma(Q)\mathrm{sub}\,P + \frac{1}{2i}\{\sigma(P),\sigma(Q)\}.$$

(Here $\{\cdot,\cdot\}$ is the Poisson bracket.)

2) Let $k \in \mathbb{N}$. Find an expression for $\mathrm{sub}(P^k)$ where $P^k = P \circ P \circ \ldots \circ P$, k times. What is the degree of homogeneity of $\mathrm{sub}(P^k)$?

3) Show that $P \equiv Op(a) \bmod L^{-\infty}$, $a \in S_{cl}^m$, where Op denotes the Weyl quantization (Exercise 3.1), and $p(x,\xi) \sim \sum \frac{1}{\alpha!} 2^{-|\alpha|} \partial_\xi^\alpha D_x^\alpha a(x,\xi)$. In particular, if $a \sim a_m + a_{m-1} + \ldots$ with a_k homogeneous of degree k, show that $a_{m-1} = \mathrm{sub}\,P$.

Exercise 5.7

Let $\Omega \subset \mathbb{R}^2$ be open and let $\varphi \in C^\infty(\Omega;\mathbb{R})$ be a solution to the Hamilton–Jacobi equation

$$(*) \qquad (\partial_{x_1}\varphi)^2 + (\partial_{x_2}\varphi)^2 - 1 = 0.$$

a) Show that the corresponding Lagrangian manifold $\Lambda_\varphi = \{(x,\partial_x\varphi(x))\,;\,x \in \Omega\}$ is a union of line segments. Deduce that the integral curves of the gradient vector field $\partial_{x_1}\varphi(x)\,\partial_{x_1} + \partial_{x_2}\varphi(x)\,\partial_{x_2}$ are line segments.

b) Let $\Gamma_t \subset \Omega$ denote the level curve defined by $\varphi(x) = t$. Let $x_0 \in \Omega$. Show that for all x in some neighborhood of x_0, we have $|\varphi(x) - \varphi(x_0)| = d(x,\Gamma_{\varphi(x_0)})$, where d is the Euclidean distance.

Now let $]a,b[\ni t \mapsto \gamma(t) \in \mathbb{R}^2$ be a smooth curve with γ injective, $\|\dot\gamma(t)\| = 1$, and let $x_0 = \gamma(t_0)$ be a point on the (image of the) curve.

c) Show that in some sufficiently small disk Ω there are precisely two smooth (real) solutions of $(*)$ satisfying $\varphi\,|_\gamma = 0$, and that for one of the two solutions we have

$$(**)\quad \Lambda_\varphi = \{(\gamma(t) + sn(t); n(t))\,;\,(t,s) \in \text{ a suitable neighborhood of } (t_0,0)\}$$

where $n(t) = (-\dot\gamma_2(t),\dot\gamma_1(t))$ is one of the normals of γ at $\gamma(t)$.

d) If we assume $\ddot\gamma(t_0) \neq 0$ (recall that this vector describes the curvature of γ at $\gamma(t_0)$ and is orthogonal to $\dot\gamma(t_0)$), and extend the definition of Λ_φ in $(**)$ by allowing s to vary in $\mathbb{R}$, show that the projection $\Lambda_\varphi \ni (x,\xi) \mapsto x$ is not a local diffeomorphism near $(\gamma(t_0) + s_0 n(t_0); n(t_0))$ for precisely one value of s_0. Determine this value and describe the corresponding base point $\gamma(t_0) + s_0 n(t_0)$ in terms of the radius of curvature of γ at $\gamma(t_0)$.

e) In the case $\Omega = \mathbb{R}^2$, show that the only smooth real-valued solutions of (*) are of the form

$$\varphi(x) = x \cdot \omega_0 + t_0 \quad \text{for} \quad \omega_0 \in \mathbb{R}^2, \quad \|\omega_0\| = 1, \quad t_0 \in \mathbb{R}.$$

Notes

The presentation has been inspired by that of Duistermaat [D]. Symplectic geometry is closely related to classical mechanics. See also Sternberg [St].

6 The strictly hyperbolic Cauchy problem − construction of a parametrix

Let P be a differential operator of order m with C^∞ coefficients on an open set $\Omega \subset\subset \mathbb{R}^n$. Let $p(x,\xi)$ be the principal symbol, a homogeneous polynomial of degree m in the ξ variables. We say that P is strictly hyperbolic, with respect to the family of hyperplanes $x_n = \text{const.}$, if the equation $p(x,\xi) = 0$ has precisely m *real* and *distinct* roots for every $(x,\xi') \in \Omega \times \dot{\mathbb{R}}^{n-1}$. After multiplying P to the left (and hence also p) by a non-vanishing function $f \in C^\infty(\Omega)$, we may assume that

$$(6.1) \qquad p = \prod_{j=1}^{m} (\xi_n - \lambda_j(x,\xi'))$$

where λ_j are the roots, and it is easy to show that $\lambda_j \in C^\infty(\Omega \times \dot{\mathbb{R}}^{n-1})$. Also,

$$(6.2) \qquad P = D_{x_n}^m + A_1(x, D_{x'}) D_{x_n}^{m-1} + \ldots + A_m(x, D_{x'}),$$

where A_j is a differential operator of order $\leq j$.

Let $\omega = \{x' \in \mathbb{R}^{n-1} \ ; \ (x',0) \in \Omega\}$, and assume that Ω is of the form $\bigcup_{x' \in \omega} \{y \in \mathbb{R}^n \ ; \ |y' - x'| + \lambda(x')|y_n| < \frac{1}{2} d(x')\}$, where $d(x')$ is the distance from x' to $C\omega = \mathbb{R}^{n-1} \setminus \omega$ and where $\lambda(x') > 0$ for every $x' \in \omega$. (For every positive function λ, this is an open set with $\Omega \cap \{x_n = 0\} = \omega \times \{x_n = 0\}$.) We say that Ω is "sufficiently flat" if $\lambda(x') \geq \mu(x')$ for some "sufficiently large" function $\mu(x') > 0$. In this chapter we shall construct solutions modulo C^∞ to the Cauchy problem fairly explicitly. Such approximate solutions are of great interest in spectral theory and can also be used in proving existence results (via the so-called Duhamel's principle). A more efficient way of getting existence and uniqueness is by use of weighted L^2-inequalities. In this book we simply recall one such result.

Theorem 6.1 *Let P be as above. After replacing Ω by a sufficiently flat open set with the same intersection $\omega \times \{x_n = 0\}$ with $x_n = 0$, we have :*

(A) *For every $v \in \mathcal{D}'(\Omega)$, $(H_{\text{loc}}^s(\Omega), C^\infty(\Omega),)$ there exists $u \in \mathcal{D}'(\Omega)$, $(H_{\text{loc}}^{s+m-1}(\Omega), C^\infty(\Omega),)$ such that $Pu = v$.*

(B) *For all $v_0, \ldots, v_{m-1} \in \mathcal{D}'(\omega)$ there exists $u \in \mathcal{D}'(\Omega)$ such that :*

  1) $u|_{I \times \tilde{\omega}} \in C^\infty(I; \mathcal{D}'(\tilde{\omega}))$ *for all open $\tilde{\omega} \subset\subset \omega$ and open intervals $I \subset \mathbb{R}$ such that $\tilde{\omega} \times I \subset\subset \Omega$,*

  2) $Pu = 0$, $D_{x_n}^j u|_{x_n = 0} = v_j$, $0 \leq j \leq m - 1$.

(C) The solution u in (B) is unique and if $v_j \in C^\infty(\tilde{\omega})$, $0 \leq j \leq m-1$, then $u \in C^\infty(\Omega)$.

Here $C^\infty(I; \mathcal{D}'(\tilde{\omega}))$ denotes the space of C^∞ functions (for the weak topology on $\mathcal{D}'(\tilde{\omega}))$: $I \to \mathcal{D}'(\tilde{\omega})$. This space can be viewed as a subspace of $\mathcal{D}'(\tilde{\omega} \times I)$ (using the Banach–Steinhaus theorem).

We shall construct u in (B) modulo $C^\infty(\Omega)$. For every root $\lambda = \lambda_\nu$, we shall first construct a certain operator $E_\nu : \mathcal{E}'(\omega) \to \mathcal{D}'(\Omega)$, such that $P \circ E_\nu \in I^{-\infty}(\Omega \times \omega)$. Here we let $I^{-\infty}(\Omega \times \omega)$ denote the space of operators $C_0^\infty(\omega) \to \mathcal{D}'(\Omega)$ with a C^∞ distribution kernel (or equivalently the space of regularizing operators $\mathcal{E}'(\omega) \to C^\infty(\Omega)$).

We try:

$$(6.3) \quad E_\nu u(x) = \int e^{i\varphi_\nu(x,\eta')} a_\nu(x,\eta') \hat{u}(\eta') \frac{d\eta'}{(2\pi)^{n-1}}$$

$$= \iint e^{i(\varphi_\nu(x,\eta') - y'\eta')} a_\nu(x,\eta') u(y') dy' \frac{d\eta'}{(2\pi)^{n-1}},$$

with $\varphi = \varphi_\nu$ a suitable phase function and $a = a_\nu \in S_{1,0}^0$.

We will have $P \circ E_\nu \in I^{-\infty}$ if

$$P(e^{i\varphi_\nu(x,\eta')} a_\nu(x,\eta')) \in S^{-\infty}(\Omega \times \mathbb{R}^{n-1}).$$

Writing $e^{-i\varphi} D_{x_j} e^{i\varphi} = D_{x_j} + \partial_{x_j}\varphi$ and $e^{-i\varphi} D_x^\alpha e^{i\varphi} = (D_{x_1} + \partial_{x_1}\varphi)^{\alpha_1} \circ \ldots \circ (D_{x_n} + \partial_{x_n}\varphi)^{\alpha_n}$, we obtain

$$(6.4) \quad P\left(e^{i\varphi(x,\eta')} a(x,\eta')\right) = e^{i\varphi(x,\eta')} b(x,\eta'),$$

where $b \in S_{1,0}^m(\Omega \times \mathbb{R}^{n-1})$ (after introducing a cut-off, we may assume that all symbols vanish for sufficiently small $|\eta'|$). Modulo $S^{m-1}(\Omega \times \mathbb{R}^{n-1})$ (where for convenience we write $S^k = S_{1,0}^k$) we get

$$b \equiv p(x, \varphi'_x) a(x, \eta').$$

Our aim is to find a non-vanishing a with $b \in S^{-\infty}$ so we require that

$$(6.5) \quad p(x, \varphi'_x) = 0,$$

and we take as initial condition

$$(6.6) \quad \varphi(x', 0, \eta') = x' \cdot \eta'.$$

In (6.5) we can choose between the various roots of p. With $q = q_\nu = \xi_n - \lambda_\nu(x, \xi')$ we choose $\varphi = \varphi_\nu$ to be the solution of

$$(6.5)_\nu \quad q_\nu(x, \varphi'_x) = 0 \quad \text{(eikonal equation)}.$$

It follows from the discussion at the end of Chapter 5 that if Ω is sufficiently flat then $(6.5)_\nu$, (6.6) has a unique smooth solution $\varphi = \varphi_\nu(x, \eta')$ on $\Omega \times S^{n-2}$; and we can extend φ_ν to a smooth solution on $\Omega \times \dot{\mathbb{R}}^{n-1}$ which is positively homogeneous of degree 1 in η', by putting $\varphi_\nu(x, \eta') = |\eta'| \varphi_\nu \left(x, \frac{\eta'}{|\eta'|}\right)$.

For $a \in S^k(\Omega \times \mathbb{R}^{n-1})$ (vanishing for small $|\eta'|$), we get
$$P(e^{i\varphi(x,\eta')} a(x, \eta')) = e^{i\varphi(x,\eta')} b(x, \eta'),$$
where $b \in S^{k+m-1}(\Omega \times \mathbb{R}^{n-1})$ and
$$b \equiv \sum_{j=1}^{n} p^j(x, \varphi'_x) D_{x_j} a + f_{\varphi, P}(x, \eta') a \stackrel{\text{def}}{=} L(a) \bmod S^{k+m-2}.$$

Here $f_{\varphi, P}$ is C^∞ and positively homogeneous of degree $m - 1$ in η'.

We now construct $a_\nu = a \sim a^k + a^{k-1} + a^{k-2} + \ldots \in S^k_{\text{cl}}(\Omega \times \mathbb{R}^{n-1})$ with a^j positively homogeneous of degree j in the following way: first we construct a^k satisfying

(6.7) $\quad \begin{cases} L(a^k) = 0 & \text{(first transport equation)} \\ a^k|_{x_n=0} = c^k, \end{cases}$

where $c^k \in C^\infty(\omega \times \dot{\mathbb{R}}^{n-1})$ is positively homogeneous of degree k in η'. In order to solve (6.7) we first take $|\eta'| = 1$, and notice that $\sum p^{(j)}(x, \varphi'_x) \frac{\partial}{\partial x_j}$ is a vector field such that the coefficient $p^{(n)}$ of $\frac{\partial}{\partial x_n}$ is non-vanishing. Hence (6.7) can be solved along each integral curve of this field and we get a solution $a^k(x, \eta') \in C^\infty(\Omega \times S^{n-2})$, which can be extended by homogeneity since the coefficients of L are positively homogeneous in η' of degree $m - 1$.

We find $P(e^{i\varphi} a_k) = e^{i\varphi}(b^{m+k-2} +$ an element of $S^{m+k-3})$, where b^{m+k-2} is positively homogeneous in η' of degree $m + k - 2$. We then look for a^{k-1}, positively homogeneous of degree $k - 1$, such that

$\quad \begin{cases} L(a^{k-1}) = -b^{m+k-2} & \text{(second transport equation)} \\ a^{k-1}|_{x_n=0} = c^{k-1}, \end{cases}$

where $c^{k-1} \in C^\infty(\omega \times \dot{\mathbb{R}}^{n-1})$ is positively homogeneous of degre $k - 1$ in η'. This can be solved the same way, and by iteration we find $a \in S^k_{\text{cl}}(\Omega \times \mathbb{R}^{n-1})$ by solving
$$P(e^{i\varphi} a) \in S^{-\infty}(\Omega \times \mathbb{R}^{n-1}), \quad a|_{x_n=0} = c,$$
for any given $c \in S^k_{\text{cl}}$ which vanishes for small $|\eta'|$.

6 The strictly hyperbolic Cauchy problem

With a_ν and φ_ν as above we define the operators $E_1, \ldots, E_m$ as in (6.3). The functions $\Phi_\nu = \varphi_\nu(x, \eta') - y'\eta'$ are phase functions and we have both $\dfrac{\partial \Phi_\nu}{\partial y'} \neq 0$ and $\dfrac{\partial \Phi_\nu}{\partial x'} \neq 0$ when $\eta' \neq 0$. Hence,

$$E_\nu : C_0^\infty(\omega) \to C^\infty(\Omega)$$
$$\mathcal{E}'(\omega) \to \mathcal{D}'(\Omega),$$

and we even have that $E_\nu u|_{\tilde{\omega} \times I} \in C^\infty(I; \mathcal{D}'(\tilde{\omega}))$, for $u \in \mathcal{E}'(\omega)$, if $\tilde{\omega} \times I \subset\subset \Omega$. Moreover, $P \circ E_\nu \in I^{-\infty}(\Omega \times \omega)$, $E_\nu u|_{x_n = 0} = C_\nu u$, where C_ν is the pseudodifferential operator with symbol $a_\nu|_{x_n=0}$.

Before continuing the discussion of the Cauchy problem, we study sing supp K_{E_ν}. For $y' \in \omega$, $\eta' \in S^{n-1}$, $\nu \in \{1, \ldots, m\}$, let $\gamma_{y',\eta',\nu}$ be the maximal integral curve in Ω of $\sum q_\nu^{(j)}(x, \varphi'_{\nu x}) \dfrac{\partial}{\partial x_j}$ which passes through $(y', 0)$. In other words, it is the projection in Ω of the integral curve of H_p, which passes through $(y', 0; \eta', \lambda_\nu(y', 0, \eta'))$.

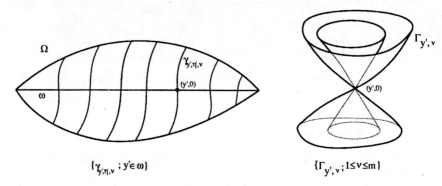

Figure 1

We have $\dfrac{\partial \Phi_\nu}{\partial \eta'} = 0$ if and only if $\dfrac{\partial \varphi_\nu}{\partial \eta'}(x, \eta') = y'$. As noticed at the end of Chapter 5, $\dfrac{\partial \varphi_\nu}{\partial \eta'}$ is constant $= y'$ on $\gamma_{y',\eta',\nu}$, so if we put $\Gamma_{y',\nu} = \bigcup_{|\eta'|=1} \gamma_{y',\eta',\nu}$, we get

$$\text{sing supp } K_{E_\nu} \subset \{(x, y') \in \Omega \times \omega ; x \in \Gamma_{y',\nu}\}.$$

For $0 \leq j \leq m - 1$, we now try to solve (approximately) the Cauchy problem

$$Pu = 0$$
$$D_{x_n}^\nu u|_{x_n=0} = \delta_{\nu,j} v, \qquad 0 \leq \nu \leq m-1$$

(where $\delta_{\nu,j}$ is Kronecker's delta : $\delta_{j,k} = 1$ for $j = k$ and 0 otherwise).

We look for u of the form

$$u = F_j v = \sum_{k=1}^m E_{j,k} v,$$

with

$$E_{j,k}v(x) = \frac{1}{(2\pi)^n} \int e^{i\varphi_k(x,\eta')} a_{j,k}(x,\eta') \hat{u}(\eta') \, d\eta',$$

where $a_{j,k} \in S_{cl}^{-j}(\Omega \times \mathbb{R}^{n-1})$, $P E_{j,k} \in I^{-\infty}(\Omega \times \omega)$.

The operator $v \mapsto D_{x_n}^\nu E_{j,k}v|_{x_n=0}$ belongs to $L_{cl}^{\nu-j}(\omega)$ and if $a_{j,k}^{-j}$ is the leading part of $a_{j,k}$, then the homogeneous principal symbol of this operator is

$$(\lambda_k(x',0,\xi'))^\nu a_{j,k}^{-j}(x',0,\xi').$$

Hence $v \mapsto D_{x_n}^\nu F_j v|_{x_n=0}$ belongs to $L_{cl}^{\nu-j}(\Omega)$ and the homogeneous principal symbol is

$$\sum_{k=1}^m (\lambda_k(x',0,\xi'))^\nu a_{j,k}^{-j}(x',0,\xi').$$

We can find unique smooth $a_{j,k}^{-j}(x',0,\xi')$, positively homogeneous of degree $-j$, such that

(6.8) $$\sum_{k=1}^m (\lambda_k(x',0,\xi'))^\nu a_{j,k}^{-j}(x',0,\xi') = \delta_{\nu,j}, \quad 0 \leq \nu \leq m-1, \ \xi' \neq 0.$$

Indeed the Van der Monde determinant

$$\det \begin{pmatrix} 1 & 1 & \cdots & 1 \\ \lambda_1 & \lambda_2 & & \lambda_m \\ \lambda_1^2 & \lambda_2^2 & & \lambda_m^2 \\ \vdots & \vdots & & \vdots \\ \lambda_1^{m-1} & \lambda_2^{m-1} & & \lambda_m^{m-1} \end{pmatrix}$$

does not vanish since the λ_j are pairwise different. The first transport equations (resulting from the requirement that $P \circ E_{j,k} \in I^{-\infty}$) now determine $a_{j,k}^{-j}(x,\xi')$ for all $x \in \Omega$. Writing $a_{j,k} \sim a_{j,k}^{-j} + a_{j,k}^{-j-1} + \ldots$ we next determine $a_{j,k}^{-j-1}(x',0,\xi')$, by solving a system of the type (6.8) with the same coefficients to the left, but with new expressions to the right; the second transport equations then determine $a_{j,k}^{-j-1}(x,\xi')$. This procedure can be repeated indefinitely and we get :

Theorem 6.2 Let P, Ω be as above. After replacing Ω by a sufficiently flat open set (still denoted by Ω) which has the same intersection $\omega \times \{0\}$

with $x_n = 0$, we can construct operators

$$F_j : \mathcal{E}'(\omega) \to \mathcal{D}'(\Omega), \qquad j = 0, 1, \ldots, m-1$$
$$C_0^\infty(\omega) \to C^\infty(\Omega)$$

of the form $F_j v(x) = \sum_{k=1}^{m} \int e^{i\varphi_k(x,\eta')} a_{j,k}(x,\eta') \hat{u}(\eta') d\eta'$ with φ_ν given by $(6.5)_\nu$, (6.6) and $a_{j,k} \in S_{cl}^{-j}(\Omega \times \mathbb{R}^{n-1})$ such that $PF_j \in I^{-\infty}(\Omega \times \omega)$ and $D_{x_n}^\nu F_j v|_{x_n=0} = \delta_{j,\nu} v$, $0 \le \nu \le m-1$. Moreover, sing supp $K_{F_j} \subset \bigcup_{k=1}^{m} \{(x,y') \,;\, x \in \Gamma_{y',k}\}$, with $\Gamma_{y',k}$ defined above.

Exercises

Exercise 6.1

Let $x = (x_1, x_2) \in \mathbb{R}^2$. Let

$$P = (x_2 D_{x_2})^2 - (D_{x_1})^2 + f(x) D_{x_1} + g(x) D_{x_2} + h(x)$$

where $f, g, h \in C^\infty(\mathbb{R}^2)$.

1) Show that P is strictly hyperbolic with respect to $x_2 = \alpha$, $0 < \alpha < +\infty$.

2) We want to solve the approximate Cauchy problem in $\Omega = \{(x_1, x_2) \,;\, x_2 > 0\}$:

$$\begin{cases} Pu \in C^\infty(\mathbb{R}^2) \\ u|_{x_2=1} = u_0 \qquad D_{x_2} u|_{x_2=1} = u_1 \end{cases}$$

where $u_0, u_1 \in \mathcal{E}'(\mathbb{R}_{x_1})$.

3) Determine explicitly two solutions $\varphi = \varphi_\pm(x, \eta) \in C^\infty(\Omega \times \mathbb{R})$ of

$$p(x, \varphi'_x) = 0 \qquad \varphi|_{x_2=1} = x_1 \eta_1.$$

Here $p = (x_2 \xi_2)^2 - \xi_1^2$.

4) Show that if $a \in S_{1,0}^m(\Omega \times \mathbb{R})$, then the operators $A_\pm$ given by

$$A_\pm v = \frac{1}{2\pi} \iint e^{i(\varphi_\pm(x,\eta_1) - y_1 \eta_1)} a(\dot{x}, \eta_1) v(y_1) \, dy_1 \, d\eta_1$$

are well defined $C_0^\infty(\mathbb{R}) \to C^\infty(\Omega)$ and $\mathcal{E}'(\mathbb{R}) \to \mathcal{D}'(\Omega)$.

5) Let $K_{A_\pm}(x, y_1) \in \mathcal{D}'(\Omega \times \mathbb{R})$ be the distribution kernel of $A_\pm$. Determine $\operatorname{sing\,supp} K_{A_\pm}$.

6) Show that if $v \in \mathcal{E}'(\mathbb{R})$,
$$\operatorname{sing\,supp}(A_\pm v) \subset \{x \in \Omega\,;\, x_1 \pm \log x_2 \in \operatorname{sing\,supp} v\}.$$

7) Show how to construct a solution to the approximate Cauchy problem.

Exercise 6.2 (Duhamel's principle)

Let M be a C^∞ compact manifold and let
$$P = D_t^m + A_1(x, D_x)\, D_t^{m-1} + \ldots + A_{m-1}(x, D_x)\, D_t + A_m(x, D_x),$$
where $A_j(x, D_x)$ is a differential operator of order j with C^∞ coefficients on M and $t \in \mathbb{R}$.

1) Give a suitable definition of strict hyperbolicity for P.
We assume from now on that P is strictly hyperbolic.

2) Construct a continuous operator $E(t): C^\infty(M) \to C^\infty(]-\varepsilon_0, \varepsilon_0[\times M)$ and $\mathcal{D}'(M) \to C^\infty(]-\varepsilon_0, \varepsilon_0[\,;\, \mathcal{D}'(M))$, such that $E(t)v(x) = u(t, x)$ solves
$$\begin{cases} PEv = Kv \\ D_t^j Ev|_{t=0} = 0 & j = 0, 1, \ldots, m-2 \\ D_t^{m-1} Ev|_{t=0} = v \end{cases}$$
with $K \in I^{-\infty}(]-\varepsilon_0, \varepsilon_0[\times M \times M)$.

3) If $v \in C^\infty([0, \varepsilon_0[\times M)$ define
$$Fv(t, x) = i\int_0^t E(t-s)\,v(s, \cdot)\,ds$$
$$\mathcal{K}v(t, x) = i\int_0^t K(t-s)\,v(s, \cdot)\,ds.$$

Show that
$$\begin{cases} PFv = v + \mathcal{K}v \\ D_t^j Fv|_{t=0} = 0 & j = 0, 1\ldots, m-1. \end{cases}$$

4) Show that $\mathcal{K}$ is continuous $C^\infty([0,\varepsilon_0[\times M) \to C^\infty([0,\varepsilon_0[\times M)$ and $C^k([0,\varepsilon_0[\,;\,H^s(M)) \to C^k([0,\varepsilon_0[\,;\,H^s(M))$.

Show that $I + \mathcal{K}$ has an inverse of the form $I + \mathcal{L}$ where

$$\mathcal{L}v = \int_0^t L(t-s)\,v(s,\cdot)\,ds$$

has a kernel $L(t-s,x,y) \in C^\infty([0,\varepsilon[\times M \times M)$. (Study the convergence of the Neumann series $I - \mathcal{K} + \mathcal{K}^2 - \mathcal{K}^3 + \ldots$.)

Finally show that $F_{\text{cor}} = F(I + \mathcal{L})$ solves the Cauchy problem

$$\begin{cases} P F_{\text{cor}} v = v \\ D_t^j F_{\text{cor}} v|_{t=0} = 0. \end{cases}$$

5) Show how to solve the Cauchy problem with non-vanishing data

$$\begin{cases} Pu = v \\ D_t^j u|_{t=0} = v_j \end{cases} \qquad j = 0, \ldots, m-1.$$

Exercise 6.3

Let $\varphi \in C^\infty(\mathbb{R}^2\,;\,\mathbb{R})$ with $\|\nabla\varphi(x)\| = 1$ for all x with $\varphi(x) = 0$. Put

$$v(x) = \int_0^\infty e^{i\lambda\varphi(x)}\,a(x,\lambda)\,d\lambda \in \mathcal{D}'(\mathbb{R}^2),$$

where $a \in S^0_{\text{cl}}(\mathbb{R}^2 \times \mathbb{R})$. Let $x_0 \in \mathbb{R}^2$ be a point where φ vanishes. Construct

$$u(t,x) = \sum_{j \in \{+,-\}} \int_0^\infty e^{i\lambda\varphi_j(t,x)}\,a_j(t,x,\lambda)\,d\lambda,$$

defined for (t,x) in a neighborhood Ω of $(0,x_0)$, such that

$$(D_t^2 - (D_{x_1}^2 + D_{x_2}^2))u \in C^\infty(\Omega), \quad u(0,x) = v(x), \quad D_t u(0,x) = 0,$$

by first solving the eikonal equation

$$(\partial_t \varphi_\pm)^2 - (\partial_x \varphi_\pm)^2 = 0,$$

and then a number of transport equations for $a_\pm$. Show that sing supp $u(t,\cdot) \subset \{x \pm t\,\nabla\varphi(x)\,;\,\varphi(x) = 0\}$.

Notes

The construction in this chapter of asymptotic solutions for general strictly hyperbolic problems was given by Lax [L], and forms an example of the classical WKB method. It plays a central role in spectral theory, and one of the main motivations for the global theory of Fourier integral operators (see Chapter 11) was to understand the global nature of the solutions to strictly hyperbolic Cauchy problems.

7 The wavefront set (singular spectrum) of a distribution

If $X \subset \mathbb{R}^n$ is open, $x_0 \in X$, $u \in \mathcal{D}'(X)$, then $x_0 \notin \operatorname{sing\,supp}(u)$ if and only if there exists $\varphi \in C_0^\infty(X)$ with $\varphi(x_0) \neq 0$ such that $\varphi u \in C^\infty(X)$. Using pseudodifferential operators instead of cut-off functions, we get a refinement of the singular support, namely the wavefront set, $WF(u) \subset T^*X \backslash 0$. All pseudodifferential operators in this chapter, except in the proof of Proposition 7.4, will be assumed properly supported.

Definition 7.1 Let $u \in \mathcal{D}'(X)$, $(x_0, \xi_0) \in T^*X\backslash 0$. We say that u is C^∞ (microlocally) near (x_0, ξ_0) if there exists $A \in L_{1,0}^m(X)$ non-characteristic at (x_0, ξ_0), such that $Au \in C^\infty(X)$. We let $WF(u)$ be the set of points in $T^*X\backslash 0$, where u is not C^∞.

The set of points in $T^*X\backslash 0$ where u is C^∞ is an open conic set, so $WF(u)$ is a closed conic set in $T^*X\backslash 0$.

If $V \subset T^*X\backslash 0$, we let $S_{\rho,\delta}^m(V) = \big\{ a \in C^\infty(V)$; for every open conic set $W \subset\subset V$ and all $\varepsilon > 0$, $\alpha, \beta \in \mathbb{N}^n$, there exists $C = C_{\varepsilon,\alpha,\beta,W} > 0$ such that $|\partial_x^\alpha \partial_\xi^\beta a| \leq C(1+|\xi|)^{m-\rho|\beta|+\delta|\alpha|}$ for $(x,\xi) \in W$ with $|\xi| \geq \varepsilon \big\}$.

Here $W \subset\subset V$ means that $W \cap (X \times S^{n-1}) \subset\subset V \cap (X \times S^{n-1})$. Similarly we define $S^{-\infty}(V)$. The reader may verify easily that the general results for symbols in Chapter 1 remain valid.

If $A \in L_{1,0}^m(X)$, we define $WF(A)$ as the smallest closed cone $\Gamma \subset T^*X\backslash 0$ such that $\sigma_A|_{\complement\Gamma} \in S^{-\infty}(\complement\Gamma)$. The composition result for pseudodifferential operators shows that $WF(A \circ B) \subset WF(A) \cap WF(B)$. Moreover, $WF(A)$ is empty if and only if $A \in L^{-\infty}(X)$.

Lemma 7.2 If $u \in \mathcal{D}'(X)$, $A \in L_{1,0}^m(X)$, then $WF(Au) \subset WF(A) \cap WF(u)$.

Proof: Let $(x_0, \xi_0) \in (T^*X\backslash 0) \backslash (WF(A) \cap WF(u))$.

Case 1: $(x_0, \xi_0) \notin WF(A)$. Then if $B \in L_{1,0}^0(X)$ is non-characteristic at (x_0, ξ_0) with $WF(B) \cap WF(A) = \emptyset$, we have $BA \in L^{-\infty}(X)$ and hence $BAu \in C^\infty(X)$. Then $(x_0, \xi_0) \notin WF(Au)$.

Case 2: $(x_0, \xi_0) \notin WF(u)$. Let $B \in L_{1,0}^m(X)$ be non-characteristic at (x_0, ξ_0) such that $Bu \in C^\infty(X)$. Let V be an open conic neighborhood of (x_0, ξ_0) such that B is non-characteristic at every point of V. Then we can construct $c \in S_{1,0}^{-m}(V)$ such that $c \# \sigma_B = 1$ mod $S^{-\infty}(V)$ (with $\#$ defined as in Remark 3.7). Let $W \subset\subset V$ be an open conic neighborhood of (x_0, ξ_0) and let $\chi(x,\xi) \in C^\infty(T^*X)$ vanish for small $|\xi|$, be positively homogenous of degree 0 in ξ for $|\xi| \geq 1$, be equal to 1 for $(x, \xi) \in W$ and $|\xi| \geq 1$ and have its support in some cone $\subset\subset V$. Then if $C \in L_{1,0}^0(X)$ has the symbol $(\chi c) \# \sigma_B$ mod

$S^{-\infty}$, we have $Cu \in C^\infty(X)$. On the other hand $\sigma_C \equiv 1 \bmod S^{-\infty}$ in W. Let $E \in L^0_{1,0}(X)$ be non-characteristic at (x_0, ξ_0) with $WF(E) \subset W$. Then $EA \equiv EAC \bmod L^{-\infty}(X)$. Hence $EAu \equiv EA(Cu) \equiv 0 \bmod C^\infty(X)$, so $(x_0, \xi_0) \notin WF(Au)$.

□

Proposition 7.3 *Let $\Pi : T^*X \to X$ be the natural projection. Then $\Pi(WF(u)) = \operatorname{sing\,supp}(u)$.*

Proof:

1) Let $x_0 \in X \setminus \operatorname{sing\,supp}(u)$. Then there exists $\chi \in C_0^\infty(X)$ such that $\chi u \in C^\infty(X)$. Now, multiplication by χ is a pseudodifferential operator with symbol $\chi(x)$ and hence non-characteristic at every point (x_0, ξ_0) with $\xi_0 \neq 0$. Hence $\Pi(WF(u)) \subset \operatorname{sing\,supp}(u)$.

2) Let $x_0 \notin \Pi(WF(u))$. Then for every $\xi \in S^{n-1}$, there exists $A_\xi \in L^0_{1,0}(X)$, non-characteristic at (x_0, ξ) such that $A_\xi u \in C^\infty(X)$. By compactness, we can find finitely many $A_1, \ldots, A_N \in L^0_{1,0}(X)$ such that for every $\xi \in S^{n-1}$, at least one of the A_j is non-characteristic at (x_0, ξ). Then $A = \sum_1^N A_j^* A_j$ is non-characteristic at every point (x_0, ξ), $\xi \neq 0$. Since $Au \in C^\infty(X)$ we then deduce (as in Chapter 3) that u is C^∞ near x_0. Hence, $\operatorname{sing\,supp}(u) \subset \Pi(WF(u))$.

□

Proposition 7.4 *Let $u \in \mathcal{D}'(X)$, $(x_0, \xi_0) \in T^*X \setminus 0$. Then we have $(x_0, \xi_0) \notin WF(u)$ if and only if there exists $\varphi \in C_0^\infty(X)$ with $\varphi(x_0) \neq 0$ and a conic neighborhood Γ of ξ_0 in $\dot{\mathbb{R}}^n$, such that $\widehat{\varphi u}(\xi)$ is of rapid decrease in Γ. $\bigl($That is, for every $N > 0$ we have $|\widehat{\varphi u}(\xi)| \leq C_N (1 + |\xi|)^{-N}$, $\xi \in \Gamma$.$\bigr)$*

Proof: We first assume that we have φ, Γ as in the proposition. Let $\chi = \chi(\xi) \in S^0(\mathbb{R}^n)$ (with no "base variable" x) be positively homogeneous of degree 0 for $|\xi| \geq 1$ with $\chi(t\xi_0) = 1$ for $t \gg 1$. Then $Au \in C^\infty(X)$, where

$$Au(x) = \iint e^{i(x-y)\xi} \varphi(x) \varphi(y) \chi(\xi) u(y) \, dy \, \frac{d\xi}{(2\pi)^n},$$

since the inverse Fourier transform of a rapidly decreasing function is C^∞. On the other hand $A \in L^0_{1,0}(X)$ is non-characteristic at (x_0, ξ_0), so $(x_0, \xi_0) \notin WF(u)$.

We next assume that $(x_0, \xi_0) \notin WF(u)$. Let $A \in L^0(X)$ be non-characteristic at (x_0, ξ_0) and properly supported with $Au \in C^\infty(X)$. There exist

$\varphi \in C_0^\infty(X)$ with $\varphi(x_0) \neq 0$ and $\chi \in S^0(\mathbb{R}^n)$ positively homogeneous for $|\xi| \geq 1$ and with $\chi(t\xi_0) = 1$ for $t \gg 1$, such that $\chi(D)\varphi = B \circ A + R$, where $B \in L^0(X)$ and $R \in L^{-\infty}$. (Take χ and φ with sufficiently small supports and construct B roughly as the parametrix in Chapter 3.) We deduce that $\chi(D)\varphi u \in C^\infty(X)$. On the other hand $\chi(D)$ a convolution operator on $\mathbb{R}^n$: $\chi(0) = k*$, where $k \in \mathcal{E}'(\mathbb{R}^n)$ is of class $\mathcal{S}$ outside 0. Considering φu as an element of $\mathcal{E}'(\mathbb{R}^n)$ with support in X, we find that $\chi(D)\varphi u \in \mathcal{S}(\mathbb{R}^n)$. Hence $\widehat{\chi(D)\varphi u} = \chi(\xi)\widehat{\varphi u}(\xi)$ is rapidly decreasing, so $\widehat{\varphi u}(\xi)$ is rapidly decreasing in a conic neighborhood of ξ_0. □

As an application we let $\varphi(x,\theta)$ be a real-valued phase function on $X \times \dot{\mathbb{R}}^N$ and let $a \in S_{\rho,\delta}^m(X \times \mathbb{R}^N)$, with $\delta < 1$, $\rho > 0$. We put

$$\Lambda_\varphi = \{(x, \varphi'_x(x,\theta));\ \varphi'_\theta(x,\theta) = 0,\ \theta \neq 0\},$$

which is a closed conic set in $T^*X \backslash 0$. Then

$$WF(I(a,\varphi)) \subset \Lambda_\varphi.$$

In order to prove this, we let $(x_0,\xi_0) \in ((T^*X)\backslash 0)\backslash \Lambda_\varphi$. Modifying $I(a,\varphi)$ by a C^∞ function, we may assume that a has its support in an arbitrarily small closed conic neighborhood Γ of $C_\varphi = \{(x,\theta) \in X \times \dot{\mathbb{R}}^N;\ \varphi'_\theta(x,\theta) = 0\}$. If $\psi \in C_0^\infty(X)$ has its support in a sufficiently small neighborhood of x_0, we may assume that $\xi_0 \neq \varphi'_x(x,\theta) \neq 0$ for all $(x,\theta) \in (\operatorname{supp}\psi \times \dot{\mathbb{R}}^N) \cap \Gamma$. Using the homogeneity we see that there exists $C > 0$ such that

$$\left|\frac{\partial}{\partial x}(\varphi(x,\theta) - x \cdot \xi)\right| \geq \frac{1}{C}(|\theta| + |\xi|)$$

for $(x,\theta) \in (\operatorname{supp}\psi \times \dot{\mathbb{R}}^N) \cap \Gamma$ and for ξ in some sufficiently small conic neighborhood V of ξ_0. In the integral

$$\widehat{\psi I(a,\varphi)}(\xi) = \iint e^{i(\varphi(x,\theta) - x \cdot \xi)} a(x,\theta)\psi(x)\,dx\,d\theta,$$

we can make repeated integrations by parts, using the operator

$${}^tL = \frac{1}{|\varphi'_x(x,\theta) - \xi|^2} \sum_{j=1}^n \left(\frac{\partial\varphi}{\partial x_j} - \xi_j\right)D_{x_j},$$

and show that $\widehat{\psi I(a,\varphi)}(\xi)$ is rapidly decreasing in V. Choosing ψ with $\psi(x_0) \neq 0$, we deduce that $(x_0,\xi_0) \notin WF(I(a,\varphi))$, which completes the proof.

Certain operations (not possible in general) are possible for distributions under some assumptions on their wavefront sets (such as taking products and restrictions). In order to give precise statements it is convenient to introduce a topology on the space of distributions with their wavefront set in some given closed cone.

Let $\Gamma \subset T^*X\setminus 0$ be a closed cone and let $\mathcal{D}'_\Gamma(X) = \{u \in \mathcal{D}'(X); WF(u) \subset \Gamma\}$. We equip $\mathcal{D}'_\Gamma(X)$ with the topology given by all the seminorms on $\mathcal{D}'(X)$ for the weak topology : $P_\varphi(u) = |\langle \varphi, u\rangle|$, $\varphi \in C_0^\infty(X)$, as well as all seminorms of the form

$$P_{\varphi,V,N}(u) = \sup_{\xi \in V} |\widehat{\varphi u}(\xi)|(1+|\xi|)^N$$

where $N \geq 0$, $\varphi \in C_0^\infty(X)$, and $V \subset \mathbb{R}^n$ is a closed cone with (supp $\varphi \times V) \cap \Gamma = \emptyset$. That $P_{\varphi,V,N}(u)$ is finite for every $u \in \mathcal{D}'_\Gamma(X)$ requires some thought; see the remark below and Exercise 7.9.

Hence, a sequence $u_j \in \mathcal{D}'_\Gamma(X)$ converges to u in $\mathcal{D}'_\Gamma(X)$ if and only if $u_j \to u$ weakly in $\mathcal{D}'(X)$ and $P_{\varphi,V,N}(u-u_j) \to 0$ for all $P_{\varphi,V,N}$ as above.

Remark : If we have a family of $\varphi_\alpha \in C_0^\infty(X)$ and of closed cones $V_\alpha \subset \mathring{\mathbb{R}}^n$, $\alpha \in \mathcal{A}$, such that

$$(\text{supp}\,\varphi_\alpha \times V_\alpha) \cap \Gamma = \emptyset, \quad \bigcup_{\alpha \in \mathcal{A}} \{(x,\xi) \in T^*X\setminus 0;\, \varphi_\alpha(x) \neq 0,\, \xi \in \mathring{V}_\alpha\} = \complement\Gamma,$$

then the topology on $\mathcal{D}'_\Gamma(X)$ is already given by the seminorms on $\mathcal{D}'(X)$ for the weak topology and the seminorms $P_{\varphi_\alpha, V_\alpha, N}$. See Exercise 7.9.

Proposition 7.5 $C^\infty(X)$ is dense in $\mathcal{D}'_\Gamma(X)$. More precisely for every $u \in \mathcal{D}'_\Gamma(X)$ there exists a sequence $u_j \in C^\infty(X)$, $j \in \mathbb{N}$ such that $u_j \to u$ in $\mathcal{D}'_\Gamma(X)$.

Such a sequence can be found by a standard regularization argument.

We put $\mathcal{E}'_\Gamma(X) = \mathcal{D}'_\Gamma(X) \cap \mathcal{E}'(X)$ with the obvious notion of convergence. We also put $-\Gamma = \{(x,-\xi);\, (x,\xi) \in \Gamma\}$.

Proposition 7.6 Let $\Gamma_1, \Gamma_2 \subset T^*X\setminus 0$ be closed cones with $\Gamma_1 \cap (-\Gamma_2) = \emptyset$. Then the map $C^\infty(X) \times C_0^\infty(X) \ni (u,v) \mapsto \langle u,v\rangle = \int u(x)v(x)\,dx \in \mathbb{C}$ has a unique continuous (in the sense of sequences) extension : $\mathcal{D}'_{\Gamma_1}(X) \times \mathcal{E}'_{\Gamma_2}(X) \to \mathbb{C}$.

Proof: The uniqueness follows from the density of the inclusions $C^\infty \subset \mathcal{D}'_{\Gamma_1}$, $C_0^\infty \subset \mathcal{E}'_{\Gamma_2}$.

Let $u \in \mathcal{D}'_{\Gamma_1}(X)$, $v \in \mathcal{E}'_{\Gamma_2}(X)$. If $x_0 \in X$, let $\varphi \in C_0^\infty(X)$ with $\varphi(x_0) \neq 0$ satisfy $V_1 \cap -V_2 = \emptyset$, where $V_j = \{\xi \in \mathring{\mathbb{R}}^n;\, \exists x \in \text{supp}\,\varphi,\, (x,\xi) \in \Gamma_j\}$. Then we can define $\langle \varphi u, \varphi v\rangle$ by the Parseval formula

$$\langle \varphi u, \varphi v\rangle = \frac{1}{(2\pi)^n} \int \widehat{\varphi u}(\xi)\,\widehat{\varphi v}(-\xi)\,d\xi.$$

The integral converges since $\widehat{\varphi u}(\xi)$ and $\widehat{\varphi v}(-\xi)$ are of rapid decrease outside any conic neighborhood of V_1 and V_2 respectively. The continuity is also easy to establish if we notice that if $\{u_j\}$ is a weakly bounded family in $\mathcal{D}'(X)$, then by the Banach–Steinhaus theorem for Fréchet spaces, there exist $C_0 > 0$, $N_0 \geq 0$ independent of j, such that

$$|\widehat{\varphi u_j}(\xi)| \leq C_0 (1 + |\xi|)^{N_0}.$$

Moreover if $u_j \to u$ weakly in $\mathcal{D}'(X)$, then $|\widehat{\varphi u} - \widehat{\varphi u_j}(\xi)| \to 0$ locally uniformly in ξ. We now define $\langle u, v \rangle = \sum \langle \varphi_j u, \varphi_j v \rangle$ where $1 = \sum \varphi_j^2$ is a locally finite partition of unity and with each function φ_j as "φ" above. □

Proposition 7.7 *Let $X \subset \mathbb{R}^{n_X}$, $Y \subset \mathbb{R}^{n_Y}$ be open sets, and let $\Gamma_1 \subset T^*X \backslash 0$, $\Gamma_2 \subset T^*Y \backslash 0$ be closed cones. Then the map $(u, v) \mapsto u \otimes v$ from $\mathcal{D}'_{\Gamma_1}(X) \times \mathcal{D}'_{\Gamma_2}(Y)$ to $\mathcal{D}'_{(\Gamma_1 \times \Gamma_2) \cup (\Gamma_1 \times O_Y) \cup (O_X \times \Gamma_2)}(X \times Y)$ is continuous for sequences. Here $O_X = X \times \{0\}$, $O_Y = Y \times \{0\}$.*

The proof is left as an exercise. If $K \subset \mathcal{D}'(X \times Y)$, we denote by the same letter the corresponding operator $C_0^\infty(Y) \to \mathcal{D}'(X)$. A typical point of $T^*(X \times Y)$ will be denoted by $(x, \xi; y, \eta)$ rather than $((x, y), (\xi, \eta))$. Put :

$$\begin{aligned} WF'(K) &= \{(x, \xi; y, -\eta) \in T^*(X \times Y) \backslash 0; (x, \xi; y, \eta) \in WF(K)\} \\ WF'_X(K) &= \{(x, \xi) \in T^*X \backslash 0; \exists y \in Y \text{ with } (x, \xi; y, 0) \in WF'(K)\} \\ WF'_Y(K) &= \{(y, \eta) \in T^*Y \backslash 0; \exists x \in X \text{ with } (x, 0; y, \eta) \in WF'(K)\}. \end{aligned}$$

If we consider $WF'(K)$ as a relation $T^*Y \to T^*X$, then $WF'_X(K)$ is the image of O_Y and $WF'_Y(K)$ is the inverse image of O_X. The next result says essentially that if $u \in \mathcal{E}'(Y)$ and $WF(u) \cap WF'_Y(K) = \emptyset$, then Ku is "well defined" in $\mathcal{D}'(X)$ and

$$WF(Ku) \subset WF'(K)(WF(u)) \cup WF'_X(K).$$

We formulate the result only in the case when K is properly supported. Then $WF'_X(K)$ and $WF'_Y(K)$ are closed.

Theorem 7.8 *Let K be properly supported and let $\Gamma \subset T^*Y \backslash 0$ be a closed cone with $WF'_Y(K) \cap \Gamma = \emptyset$. Then $\tilde{\Gamma} = WF'(K)(\Gamma) \cup WF'_X(K)$ is closed in $T^*X \backslash 0$ and K has a (unique) continuous extension for sequences $\mathcal{D}'_\Gamma(Y) \to \mathcal{D}'_{\tilde{\Gamma}}(X)$.*

Proof: To show that $\tilde{\Gamma}$ is closed, it suffices to show that $\overline{WF'(K)(\Gamma)} \subset \tilde{\Gamma}$ (where the bar indicates closure in $T^*X \backslash 0$). Let $(x_j, \xi_j) \in WF'(K)(\Gamma)$ be a sequence such that $(x_j, \xi_j) \to (x_0, \xi_0) \in T^*X \backslash 0$. Then $\exists (y_j, \eta_j) \in \Gamma$ with $(x_j, \xi_j; y_j, \eta_j) \in WF'(K)$. After passing to a subsequence, we have only three possibilities :

1) $(y_j, \eta_j) \to (y_0, 0), y_0 \in X$. Then $(x_0, \xi_0) \in WF'_X(K) \subset \tilde{\Gamma}$.

2) $(y_j, \eta_j) \to (y_0, \eta_0)$, $y_0 \in X, \eta_0 \neq 0$. Then $(y_0, \eta_0) \in \Gamma$, so $(x_0, \xi_0) \in WF'(K)(\Gamma) \subset \tilde{\Gamma}$.

3) $|\eta_j| \to \infty$, $\left(y_j, \dfrac{\eta_j}{|\eta_j|}\right) \to (y_0, \eta_0)$, $y_0 \in X$. Then $\left(x_j, \dfrac{\xi_j}{|\eta_j|}; y_j, \dfrac{\eta_j}{|\eta_j|}\right) \to (x_0, 0; y_0, \eta_0) \in WF'(K)$, and we get $(y_0, \eta_0) \in WF'_Y(K) \cap \Gamma$, which is impossible by assumption.

Let $u \in \mathcal{D}'_\Gamma(Y)$, $v \in C_0^\infty(X)$. Then $WF(v \otimes u) \subset O_X \times \Gamma$ and since $(O_X \times (-\Gamma)) \cap WF(K) = \emptyset$ by assumption, we can define $\langle K, v \otimes u \rangle$ by Proposition 7.6. For a fixed $u \in \mathcal{D}'_\Gamma(Y)$, the mapping $C_0^\infty(X) \ni v \mapsto v \otimes u \in \mathcal{D}'_{O_X \times \Gamma}(X \times Y)$ is sequentially continuous. Hence $\langle Ku, v \rangle_X = \langle K, v \otimes u \rangle_{X \times Y}$ defines a distribution $Ku \in \mathcal{D}'(X)$. For a fixed $v \in C_0^\infty(X)$, the map $\mathcal{D}'_\Gamma(Y) \ni u \mapsto v \otimes u \in \mathcal{D}'_{O_X \times \Gamma}(X \times Y)$ is sequentially continuous. Hence $K : \mathcal{D}'_\Gamma(Y) \to \mathcal{D}'(X)$ is sequentially continuous.

In order to estimate $WF(Ku)$, we take $\varphi \in C_0^\infty(X)$ and a closed cone $V_1 \subset \mathbb{R}^{n_x}$ such that $(\operatorname{supp}\varphi \times V_1) \cap \tilde{\Gamma} = \emptyset$. For $\xi \in V_1$, $N > 0$, we put $v_{N,\xi}(x) = \varphi(x) e^{-ix\xi}(1 + |\xi|)^N$. Then $v_{N,\xi} \to 0$ in $\mathcal{D}'_{-\Gamma_1}$ when $\xi \to \infty$ in V_1. Here we define Γ_1 to be $\operatorname{supp}\varphi \times V_1$. Hence for $u \in \mathcal{D}'_\Gamma$, $v_{N,\xi} \otimes u \to 0$ in $\mathcal{D}'_{(-\Gamma_1 \times \Gamma) \cup (-\Gamma_1 \times O_Y) \cup (O_X \times \Gamma)}(X \times Y)$. Now,

$$WF(K) \cap -((-\Gamma_1 \times \Gamma) \cup (-\Gamma_1 \times O_Y) \cup (O_X \times \Gamma)) = \emptyset,$$

and hence $\langle K, v_{N,\xi} \otimes u \rangle \to 0$ when $\xi \to \infty$ in V_1. In other words, $(1 + |\xi|)^N \widehat{\varphi Ku}(\xi) \to 0$ when $\xi \to \infty$ in V_1. Hence $Ku \in \mathcal{D}'_{\tilde{\Gamma}}$.

It remains to prove the continuity $\mathcal{D}'_\Gamma \to \mathcal{D}'_{\tilde{\Gamma}}$. Let $u_j \in \mathcal{D}'_\Gamma$ be a sequence which tends to 0 in $\mathcal{D}'_\Gamma$. If $Ku_j \not\to 0$ in $\mathcal{D}'_{\tilde{\Gamma}}$ there exist φ, V_1, N as above, such that (after passing to a subsequence)

$$\sup_{\xi \in V_1} |(1 + |\xi|)^N \widehat{\varphi Ku_j}(\xi)| \geq \text{const.} > 0.$$

We can therefore find a sequence $\xi_j \in V_1$ such that

$$|(1 + |\xi_j|)^N \widehat{\varphi Ku_j}(\xi_j)| \geq \text{const.} > 0.$$

If $\{\xi_j\}$ were bounded, we would get a contradiction with the fact that $Ku_j \to 0$ in $\mathcal{D}'$ and hence we may assume that $|\xi_j| \to \infty$. Then,

$$|(1 + |\xi_j|)^N \widehat{\varphi Ku_j}(\xi_j)| = |\langle K, v_{N,\xi_j} \otimes u_j \rangle| \geq \text{const.} > 0,$$

in contradiction with the fact that

$$v_{N,\xi_j} \otimes u_j \to 0 \quad \text{in} \quad \mathcal{D}'_{(-\Gamma_1 \times \Gamma) \cup (-\Gamma_1 \times O_Y) \cup (O_X \times \Gamma)}.$$

$\square$

We leave it as an exercise to formulate and prove the analogous results for operators which are not necessarily properly supported, acting on $\mathcal{E}'_\Gamma(Y)$.

We next formulate a few consequences of Theorem 7.8. If $Au(x) = \iint e^{i\varphi(x,y,\theta)} a(x,y,\theta) u(y) \, dy \, d\theta$ is a Fourier integral operator we recover Theorem 1.17 from Theorem 7.8 and the inclusion concerning $WF'(A)$ that we established after Proposition 7.4.

Let $\mathcal{H} : X \to Y$ be a C^∞ map. For $u \in C_0^\infty(Y)$ the inverse image or pull-back $\mathcal{H}^*u \in C^\infty(X)$ is defined by

$$\mathcal{H}^*u(x) = u(\mathcal{H}(x)) = \iint e^{i(\mathcal{H}(x)-y)\eta} u(y) \, dy \, \frac{d\eta}{(2\pi)^n}.$$

Then $\mathcal{H}^*$ is a Fourier integral operator and

$$WF'(A) \subset \{(x,\xi; y,\eta) \,;\, y = \mathcal{H}(x),\, \xi = {}^t\mathcal{H}'(x)\eta,\, \eta \neq 0\}.$$

Hence $WF'_X(A) = \emptyset$ and $WF'_Y(A) = \{(y,\eta) \,;\, \exists x \in X \text{ with } y = \mathcal{H}(x) \text{ and } {}^t\mathcal{H}'(x)\eta = 0\}$. We get

Corollary 7.9 *If $u \in \mathcal{D}'(Y)$ and $WF(u) \cap \{(y,\eta) \in T^*Y\setminus 0; \exists x \in X, y = \mathcal{H}(x),\, {}^t\mathcal{H}'(x)\eta = 0\} = \emptyset$, then $\mathcal{H}^*u$ is well defined in $\mathcal{D}'(X)$ and*

$$WF(\mathcal{H}^*u) \subset \{(x,\xi) \in T^*X\setminus 0; \exists (y,\eta) \in WF(u), y = \mathcal{H}(x), \xi = {}^t\mathcal{H}'(x)\eta\}.$$

Here the term "well defined" is of course a little vague and corresponds to a statement in the style of Theorem 7.8.

Example : Let $\mathcal{H} : \mathbb{R}^{n-1} \ni x' \mapsto (x',0) \in \mathbb{R}^n$. Then $\mathcal{H}^*u = u(x',0)$ is well defined if $WF(u) \cap \{(x',0; 0,\xi_n)\} = \emptyset$, and then $WF(u(x',0)) \subset \{(x',\xi') \,;\, \exists \xi_n \in \mathbb{R} \text{ such that } (x',0; \xi',\xi_n) \in WF(u)\}$.

If $\mathcal{H} : X \to Y$ is a C^∞ map, we can define the direct image (or pushforward) $\mathcal{H}_* : \mathcal{E}'(X) \to \mathcal{E}'(Y)$ as the adjoint of $\mathcal{H}^* : C^\infty(Y) \to C^\infty(X)$. One can then study $WF'(\mathcal{H}_*)$ and $WF(\mathcal{H}_*u)$, $u \in \mathcal{E}'(X)$. See Exercise 7.10.

Theorem 7.10 *Let $X \subset \mathbb{R}^{n_X}$, $Y \subset \mathbb{R}^{n_Y}$, $Z \subset \mathbb{R}^{n_Z}$, be three open sets and let $A_1 : C_0^\infty(Y) \to \mathcal{D}'(X)$, $A_2 : C_0^\infty(Z) \to \mathcal{D}'(Y)$ be continuous linear operators with at least one of them properly supported. If $WF'_Y(A_1) \cap WF'_Y(A_2) = \emptyset$, then $A_1 \circ A_2 : C_0^\infty(Z) \to \mathcal{D}'(X)$ is a well defined continuous operator and*

$$WF'(A_1 \circ A_2) \subset (WF'_X(A_1) \times O_Z) \cup (O_X \times WF'_Z(A_2)) \cup (WF'(A_1) \circ WF'(A_2)).$$

Idea of the proof: That $A_1 \circ A_2$ is a well defined continuous operator follows from Theorem 7.8.

In order to estimate $WF'(A_1 \circ A_2)$ we notice that the kernel $K_{A_1 \circ A_2}$ can be obtained by letting the kernel

$$\mathcal{K}(x,z;y',z') = K_{A_1}(x,y') \otimes \delta(z-z')$$

act on $K_{A_2}(y',z')$:

$$K_{A_1 \circ A_2}(x,z) = \iint (K_{A_1}(x,y') \otimes \delta(z-z')) K_{A_2}(y',z')\, dy'\, dz'.$$

One can justify this formal argument by approximating K_{A_2} by a sequence of C^∞ kernels converging in $\mathcal{D}'_{WF(K_{A_2})}$. It then suffices to apply Theorem 7.8 to $\mathcal{K}$.

□

Theorem 7.11 Let $u_1, u_2 \in \mathcal{D}'(X)$ satisfy $WF(u_1) \cap -WF(u_2) = \emptyset$. Then $u_1 u_2$ is well defined in $\mathcal{D}'(X)$ and

$$WF(u_1 u_2) \subset \{(x, \xi_1 + \xi_2)\,;\, (x, \xi_j) \in (\mathrm{supp}\, u_j \times \{0\}) \cup WF(u_j)\}.$$

Proof: Let $\mathcal{H}: X \ni x \mapsto (x,x) \in X \times X$. Then we define $u_1 u_2$ as $\mathcal{H}^*(u_1 \otimes u_2)$.

□

Exercises

Exercise 7.1

Let X be an open subset of $\mathbb{R}^n$.

a) Show that if $u \in \mathcal{D}'(X)$ is real then

$$(x, \xi) \in WF(u) \implies (x, -\xi) \in WF(u).$$

b) Suppose $0 \in X$ and $x \in X \implies -x \in X$.
Show that if $u \in \mathcal{D}'(X)$ is odd or even, then

$$(0, \xi) \in WF(u) \implies (0, -\xi) \in WF(u).$$

Exercise 7.2

Let $P = (x_2 D_{x_2})^2 - D_{x_1}^2 + 2i\mu x_2 D_{x_2}$, $\mu \geq 0$.

1) Show that for every β, the distributions
$$u_\beta(x) = H(x_2) x_2^{\alpha(\beta)} e^{\beta x_1} \in \mathcal{D}'(\mathbb{R}^2)$$
with $\alpha(\beta) = \mu + (\mu^2 + \beta^2)^{1/2}$ satisfy $Pu_\beta = 0$. (Here H is the Heaviside function.)

2) Show that $WF(u_\beta) \subset \{(x_1, x_2; \xi_1, \xi_2) ; x_2 = \xi_1 = 0, \xi_2 \neq 0\}$.

3) Show that the inclusion in 2) is in fact an equality.

Exercise 7.3

Denote by 1_A the characteristic function of a subset A of $\mathbb{R}^n$: $1_A(x) = 1$, $x \in A$; $1_A(x) = 0$, $x \notin A$.
Determine $WF(u)$ if
a) $u = 1_A \in \mathcal{D}'(\mathbb{R}^n)$ with $A = \{\varphi(x) > 0\}$, $\varphi \in C^\infty(\mathbb{R}^n, \mathbb{R})$, $\varphi'(x) \neq 0$ if $\varphi(x) = 0$.
b) $u = 1_{A_1} - a\, 1_{A_2} \in \mathcal{D}'(\mathbb{R}^2)$, $a \in \mathbb{R}$,
$$A_1 = \{(x_1, x_2) ; x_1 > 0, x_2 > 0\}$$
$$A_2 = \{(x_1, x_2) ; x_1 < 0, x_2 < 0\}$$

Discuss the results according to different values of a.

Exercise 7.4

Let Ω be a bounded convex open subset of $\mathbb{R}^2$, with a C^∞ boundary $\partial \Omega = \{\gamma(t); t \in \mathbb{R}\}$, where $\gamma : \mathbb{R} \to \mathbb{R}^2$ is C^∞, 1-periodic, with $\dot{\gamma}(t) \neq 0$, $\dot{\gamma}(t) \wedge \ddot{\gamma}(t) \neq 0$, $\gamma|_{[0,1[}$ injective.
Let μ be a measure on $\mathbb{R}^2$ supported by $\partial \Omega$ given by $\mu = \gamma_*(f(t)\,dt)$, where $f \in C^\infty(\mathbb{R})$, 1-periodic.
a) Study the asymptotics of $\hat{\mu}(\xi)$ as $|\xi| \to \infty$.
b) Study the asymptotics of $\hat{\chi}(\xi)$, where χ is the characteristic function of Ω.
c) Find $WF(\mu)$.
d) Find $WF(\chi)$.

Exercise 7.5

Let $A = \{(x_1, x_2) \in \mathbb{R}^2 ; x_1^3 \geq x_2^2\}$.
Let $u = 1_A \in \mathcal{D}'(\mathbb{R}^2)$ be the characteristic function of A.
a) Choose $\chi = \chi(x_1) \in C^\infty(\mathbb{R}, \mathbb{R})$ with $\chi(x_1) = 1$ if $x_1 < \frac{1}{2}$, $\chi(x_1) = 0$ if $x_1 > 1$.

Let $\chi_\varepsilon(x_1) = \chi\left(\dfrac{x_1}{\varepsilon}\right)$, $\varepsilon > 0$.
For $a \in \mathbb{R}$ and $\lambda > 0$ show that

$$\widehat{\chi_\varepsilon u}(a\lambda, \lambda) = \dfrac{1}{i\lambda}\int_{-\infty}^{+\infty} e^{i\lambda(t^3 - at^2)} \chi_\varepsilon(t^2)\, 2t\, dt.$$

b) For $a \neq 0$ and ε small enough, find the asymptotics of $\widehat{\chi_\varepsilon u}(a\lambda, \lambda)$ as $\lambda \to +\infty$. Show that $(0,0; a, 1) \in WF(u)$ for every $a \in \mathbb{R}$.
c) Determine $WF(u)$.

Exercise 7.6

Let $u \in \mathcal{D}'(\mathbb{R}^2)$ be defined by $u(x_1, x_2) = (ix_1 + x_2^2)^{1/2}$, where we use the usual branch of the square root.
a) Show that $u = \lim_{\varepsilon \searrow 0} u_\varepsilon$ in $\mathcal{D}'$, where $u_\varepsilon(x) = (ix_1 + x_2^2 + \varepsilon)^{1/2}$. Let $L = D_{x_2} + 2ix_2 D_{x_1}$.
Show that $Lu = 0$ and deduce that

$$WF(u) \subset \{(0,0; \xi_1, 0)\,;\ \xi_1 \in \dot{\mathbb{R}}\}.$$

b) Let $\chi \in C_0^\infty(\mathbb{R}^2)$ be equal to 1 in a neighborhood of 0. Let $0 \leq \varphi \in C_0^\infty(\mathbb{R}^2)$ have its support in an open set where $\chi = 1$ and satisfy $\varphi(0) > 0$. Show that

$$\widehat{\chi u}(\xi) = \iint e^{-i(x_1\xi_1 + x_2\xi_2) - \varphi(x)\xi_1}(ix_1 + x_2^2 + \varphi(x))^{\frac{1}{2}} \chi(x)(1 - i\partial_{x_1}\varphi(x))\, dx_1 dx_2$$

by passing to a suitable x_2-dependent complex integration contour in x_1. Deduce that $\xi_1 < 0$ if $(0,0; \xi_1, 0) \in WF(u)$ and determine $WF(u)$ completely.
c) More generally, let $f \in C^\infty(X)$, X open in $\mathbb{R}^n$, and assume that $\operatorname{Im} f(x) \geq 0$, and that $f(x) = 0 \Rightarrow df(x) \neq 0$. Let $k > 0$. Noticing that $(f(x) + i\varepsilon)^{-k} = C_k \int_0^\infty e^{i(f(x) + i\varepsilon)\tau} \tau^{-1+k}\, d\tau$, $\varepsilon > 0$, show that $\lim_{\varepsilon \searrow 0}(f(x) + i\varepsilon)^{-k} \stackrel{\text{def}}{=} (f(x) + i0)^{-k}$ exists in $\mathcal{D}'(X)$, is given by an oscillatory integral and satisfies $WF((f(x) + i0)^{-k}) \subset \{(x, \lambda df(x))\,;\ f(x) = 0,\ \lambda > 0\}$. Extend the last estimate to the case $k \in \mathbb{R}$.

Exercise 7.7

Let $\Omega = \{z \in \mathbb{C}\,;\ a < \operatorname{Re} z < b,\ 0 < \operatorname{Im} z < R\}$, where $a < b$, $R > 0$. Let $u \in C^1(\Omega \cup\,]a, b[)$ (in the sense that u, $\partial_{\operatorname{Re} z} u$, $\partial_{\operatorname{Im} z} u$ exist and are continuous on $\Omega \cup\,]a, b[$) and assume that u is holomorphic in Ω. Write $u(x + i0) = u|_{]a,b[}$.
Let $\varphi \in C_0^\infty(]a, b[)$ and put $\varphi_N(z) = \chi(\operatorname{Im} z) \sum_{j=0}^{N} \dfrac{1}{j!} \varphi^{(j)}(\operatorname{Re} z)(i \operatorname{Im} z)^j$ for $N \in \mathbb{N}$, where $\chi \in C_0^\infty(]-R, R[)$, $\chi(t) = 1$ near $t = 0$.

a) Show that $\partial_{\bar{z}} \varphi_N = \mathcal{O}(|\operatorname{Im} z|^N)$, $\varphi_N(x) = \varphi(x)$ for x real. Here $\partial_{\bar{z}} = \frac{1}{2}\left(\partial_{\operatorname{Re} z} - \frac{1}{i}\partial_{\operatorname{Im} z}\right)$.

b) Show that

$$\widehat{\varphi u(x+i0)}(\xi) = \int e^{-ix\xi} \varphi(x) u(x+i0)\, dx = \iint_\Omega e^{-iz\xi} \partial_{\bar{z}} \varphi_N(z)\, u(z)\, d\bar{z} \wedge dz.$$

c) Show that for every $N \in \mathbb{N}$, $|\widehat{\varphi u(x+i0)}(\xi)| \le C_N |\xi|^{-N}$ for $\xi < 0$.
What can be said about $WF(u(x+i0))$?

d) Let $v(z)$ be holomorphic on Ω and satisfy $|v(z)| \le C_0 |\operatorname{Im} z|^{-N_0}$ for some $N_0 \in \mathbb{N}$.
Show that $v = (\partial_z)^{N_0+2} u$ for some u as above.
Deduce that $v(x+i0) = \lim_{\varepsilon \searrow 0} v(x+i\varepsilon)$ exists in $\mathcal{D}'(]a,b[)$, and show that $v(x+i0) = \partial_x^{N_0+2}(u(x+i0))$. What can be said about $WF(v(x+i0))$?

Exercise 7.8

Let E_ν be the operator introduced in (6.3), (6.5)$_\nu$. Show that

$$WF'(E_\nu) \subset \{(x,\xi; y',\eta') \in (T^*\Omega \backslash 0) \times (T^*\omega \backslash 0) \,; \\ (x,\xi) = \exp(t\, H_{q_\nu})(y', 0\,; \eta', \lambda_\nu(y',\eta')) \text{ for some } t\}.$$

Exercise 7.9

Let φ_α, V_α, $\mathcal{A}$, Γ, X be as in the remark prior to Proposition 7.5. Let $V \subset \mathbb{R}^n$ be a closed cone, $\varphi \in C_0^\infty(X)$ and assume that $(\operatorname{supp}\varphi \times V) \cap \Gamma = \emptyset$.

a) If $\xi_0 \in V$, show that there exists a finite set $\mathcal{A}_{\xi_0} \subset \mathcal{A}$, such that $\operatorname{supp}(\varphi) \times \{\xi_0\} \subset \bigcup_{\alpha \in \mathcal{A}_{\xi_0}} \{(x,\xi) \in X \times \overset{\circ}{V}_\alpha \,;\, \varphi_\alpha(x) \ne 0\}$, and functions $\chi_{\alpha,\xi_0} \in C_0^\infty(X)$, $\alpha \in \mathcal{A}_{\xi_0}$, such that $\varphi = \sum_{\alpha \in \mathcal{A}_{\xi_0}} \chi_\alpha \varphi_\alpha$. Here $\overset{\circ}{V}_\alpha$ denotes the interior of V_α.

b) Let $u \in \mathcal{D}'_\Gamma(X)$. By Fourier transforming the identity $\varphi u = \sum \chi_\alpha \varphi_\alpha u$, show that $\widehat{\varphi u}(\xi)$ is rapidly decreasing in some conic neighborhood of ξ_0. Conclude that $P_{\varphi,V,N}(u)$ is finite for every N if $\widehat{\varphi_\alpha u}(\xi)$ is rapidly decreasing in V_α, for every $\alpha \in \mathcal{A}$.

c) Let $u_j \in \mathcal{D}'_\Gamma(X)$, $j = 1,2,\ldots$ and assume that $u_j \to 0$ in $\mathcal{D}'(X)$, $j \to \infty$ and that $P_{\varphi_\alpha,V_\alpha,N}(u_j) \to 0$, $j \to \infty$ for every $\alpha \in \mathcal{A}$ and every $N \ge 0$. Show that $P_{\varphi,V,N}(u_j) \to 0$, $j \to \infty$ for every $N \ge 0$.

Exercise 7.10

Let $A : C_0^\infty(Y) \to \mathcal{D}'(X)$ be a continuous linear map, where $X \subset \mathbb{R}^{n_X}$, $Y \subset \mathbb{R}^{n_Y}$ are open. Define the "real" and complex adjoints ${}^tA$, A^* : $C_0^\infty(X) \to \mathcal{D}'(Y)$, by

$$\langle {}^tAu, v \rangle = \langle u, Av \rangle, \quad (A^*u \mid v) = (u \mid Av),$$
$$\langle u, v \rangle = \int uv, \quad (u \mid v) = \int u\bar{v}.$$

a) Show that $WF'(A^*) = \{(y,\eta\,;\,x,\xi)\,;\,(x,\xi\,;\,y,\eta) \in WF'(A)\}$

$$WF'({}^tA) = \{(y,-\eta\,;\,x,-\xi)\,;\,(x,\xi\,;\,y,\eta) \in WF'(A)\}.$$

b) Apply the result to the study of $\mathcal{H}_*$ as indicated prior to Theorem 7.10.

Notes

The notion of wavefront set is central in microlocal analysis and has been implicit in various phase space localizations where pseudodifferential operators are used as cut-off functions. The present notion is due to Hörmander (see [Hö2]). Similar notions have been developed in other versions of microlocal analysis (see Iagolnitzer-Stapp [IaSta], Sato [Sa], Sato-Kawai-Kashiwara [SaKK], Iagolnitzer [Ia], Sjöstrand [S]).

8 Propagation of singularities for operators of real principal type

All pseudodifferential operators in this chapter are assumed to be properly supported.

Let $P \in L^m_{cl}(X)$ (with X open in $\mathbb{R}^n$) and let $p \in C^\infty(T^*X\setminus 0)$ be the principal symbol, positively homogeneous of degree m. If $u \in \mathcal{D}'(X)$, $Pu \in C^\infty(X)$, we know that $WF(u) \subset p^{-1}(0) = \{(x,\xi) \in T^*X\setminus 0 \,;\, p(x,\xi) = 0\}$. A very natural problem is then to characterize the subsets of $p^{-1}(0)$ which are of the form $WF(u)$ for some $u \in \mathcal{D}'(X)$ with $Pu \in C^\infty(X)$.

Necessary conditions on such sets are results on propagation of singularities. The case when P is elliptic is trivial and the next simplest general case is that of operators of real principal type. We say that P is of real principal type if p is real-valued and $dp(x,\xi)$ and $\sum \xi_j\, dx_j$ are linearly independent for $\xi \neq 0$. (By the Euler homogeneity relations, we know that $d_\xi\, p(x,\xi) \neq 0$ when $p(x,\xi) \neq 0$, so the last property is automatically valid when $p(x,\xi) \neq 0$.)

Theorem 8.1 *Let $P \in L^m_{cl}(X)$ have the real-valued principal symbol p (positively homogeneous of degree m) and let $[a,b] \ni t \mapsto \gamma(t) \in T^*X\setminus 0$ be a bicharacteristic strip with $-\infty < a < b < \infty$ (i.e. γ is an integral curve of H_p in $p^{-1}(0)$). Let $u \in \mathcal{D}'(X)$ satisfy $\gamma([a,b]) \cap WF(Pu) = \emptyset$. Then either $\gamma([a,b]) \subset WF(u)$ or $\gamma([a,b]) \cap WF(u) = \emptyset$.*

If dp and $\xi\, dx$ are collinear at $\gamma(t_0) = (x_0,\xi_0)$ for some $t_0 \in [a,b]$, then $H_p(\gamma(t_0)) = \lambda \sum \xi_j \dfrac{\partial}{\partial \xi_j}$ for some $\lambda \in \mathbb{R}$ and it is easy to see that $\gamma(t) = (x_0, \mu(t)\,\xi_0)$, $a \le t \le b$, and since $WF(u)$ is conic by definition, the theorem is trivial in this case.

Lemma 8.2 *Let $u \in \mathcal{D}'(X)$, $\varphi \in C^\infty(X;\mathbb{R})$, $w \in C_0^\infty(X)$ and assume that $d\varphi(x) \neq 0$ for $x \in \mathrm{supp}\, w$ and that $\{(x,\varphi'_x(x))\,;\, x \in \mathrm{supp}\, w\} \cap WF(u) = \emptyset$. Then for every $N \ge 0$ there exists $C = C_N$ such that*

$$|(u \mid w\, e^{i\lambda\varphi})_{L^2}| \le C\, \lambda^{-N}, \quad \lambda \ge 1.$$

If w, φ depend continuously on a parameter α which varies in some compact set in $\mathbb{R}^k$ for some k, then we can choose the constants C_N independent of α.

Proof: We only treat the case without a parameter. Let $A \in L^0_{cl}(X)$ satisfy $WF(A-I) \cap WF(u) = \emptyset$, $WF(A) \cap \{(x,\varphi'_x)\,;\, x \in \mathrm{supp}\, w\} = \emptyset$. Then

$$(u \mid w\, e^{i\lambda\varphi})_{L^2} = (u \mid A^*(w\, e^{i\lambda\varphi}))_{L^2} + ((I-A)u \mid w\, e^{i\lambda\varphi})_{L^2}.$$

Here $(I-A)u \in C^\infty(X)$ so by integration by parts we get the rapid decrease of the second term when $\lambda \to \infty$.

After modifying A^* by a properly supported element of $L^{-\infty}(X)$, we may assume that

$$A^*(w\, e^{i\lambda\varphi})(x) = \iint e^{i(x-y)\eta + i\lambda\varphi(y)} \chi_0(x,y)\, \overline{\sigma_A(y,\eta)}\, w(y)\, dy\, \frac{d\eta}{(2\pi)^n},$$

where $\chi_0 \in C^\infty(X \times X)$ and $\mathrm{supp}\, \chi_0$ is proper. We decompose this integral in two by introducing the factors $\chi\left(\frac{1}{\lambda}(\eta - \varphi'(y))\right)$ and $1 - \chi\left(\frac{1}{\lambda}(\eta - \varphi'(y))\right)$ with $\chi \in C_0^\infty(\mathbb{R}^n)$, satisfying $0 \leq \chi \leq 1$, $\chi(\xi) = 1$ if $|\xi| \leq \frac{\varepsilon}{2}$, $\chi(\xi) = 0$ if $|\xi| \geq \varepsilon$, for some sufficiently small $\varepsilon > 0$. Then in the support of χw, we have $\partial_x^\alpha \partial_\eta^\beta \sigma_A = \mathcal{O}(\lambda^{-\infty})$ and in the support of $(1-\chi)w$, we have $|\nabla_y((x-y)\theta + \lambda\varphi)|$ of the same order of magnitude as $(1 + \lambda + |\eta|)$ and we can make integrations by parts. In this way, we prove that $\lambda \mapsto A^*(w\, e^{i\lambda\varphi}) \in C_0^\infty(X)$ is rapidly decreasing, which implies the rapid decrease of $\lambda \mapsto (u \mid A^*(w\, e^{i\lambda\varphi}))_{L^2}$. □

In order to prove the theorem it suffices to prove :

Proposition 8.3   Under the assumptions of Theorem 8.1, there exists a number $\varepsilon_0 > 0$ such that if $t, s \in [a, b]$, $|t - s| \leq \varepsilon_0$ and $\gamma(t) \notin WF(u)$, then $\gamma(s) \notin WF(u)$.

Proof: After reparametrizing γ, we may assume that $s = 0$, and after composing P to the left with an elliptic operator of order $1 - m$, we may assume that P is of order 1.

Let $w \in C_0^\infty(X)$ and $\tilde{X} \subset\subset X$ be an open set containing $\mathrm{supp}\, w$. Applying the method of Chapter 6 (and the addition about asymptotic expansions for pseudodifferential operators that we give below) we can construct $f(t, x, \eta; \lambda) \in C^\infty([-\varepsilon_0, \varepsilon_0] \times \tilde{X} \times S^{n-1} \times [1, \infty[)$, and a real-valued $\varphi(t, x, \eta) \in C^\infty([-\varepsilon_0, \varepsilon_0] \times \tilde{X} \times S^{n-1})$ such that

1) The x-space projection of $\mathrm{supp}\, f$ is compact in $\tilde{X}$, and more precisely

$$\mathrm{supp}(x \mapsto f(t, x, \eta; \lambda)) \subset \{x(t)\, ;\, (x(t), \xi(t)) = \exp t H_p(y, \eta)$$
$$\text{for some}\ (y, \eta) \in (\mathrm{supp}\, w) \times S^{n-1}\}$$

2) $f(\cdot, \cdot\cdot, \cdot\cdot\cdot\, ; \lambda)$ is uniformly bounded in C^∞ for $\lambda \geq 1$.

3) $\dfrac{\partial \varphi}{\partial t} + p(x, \varphi'_x) = 0$, $\varphi|_{t=0} = x \cdot \eta$.

4) For every $N \in \mathbb{N}$, $\lambda^N (D_t + P^*(x, D_x))(f\, e^{i\lambda\varphi})$ is uniformly bounded in $C_{t,x,\eta}^\infty$ for $\lambda \in [1, \infty[$, and $f(0, x, \eta; \lambda) = w(x)$.

The method of solving 3) implies that $\Lambda_{\varphi_{t,\eta}} = \exp(tH_p)(\Lambda_{\varphi_{0,\eta}})$, if $\varphi_{t,\eta}(x) = \varphi(t,x,\eta)$ and $\Lambda_\varphi = \{(x,\varphi'(x))\}$.

Since P^* is a pseudodifferential operator, the asymptotic expansion of $P^*(f\,e^{i\lambda\varphi})$ requires some more work, which we shall perform below.

Now we consider $F(t,w,\eta\,;\lambda) = (f\,e^{i\lambda\varphi}\,|\,u)_{L^2_x}$. Then,

$$D_t F(t,w,\eta\,;\lambda) = ((D_t + P^*)(f\,e^{i\lambda\varphi})\,|\,u) - (f\,e^{i\lambda\varphi}\,|\,Pu).$$

If $\operatorname{supp} w$ is contained in a sufficiently small neighborhood of x_0 and η is sufficiently close to ξ_0, where $(x_0,\xi_0) = \gamma(0)$ (and we assume without loss of generality that $|\xi_0| = 1$), then the geometric construction of φ in Chapter 6, Proposition 8.2 and the hypothesis on $WF(Pu)$ imply that

$$(f\,e^{i\lambda\varphi}, Pu) = \mathcal{O}(\lambda^{-\infty}), \quad \lambda \geq 1,$$

uniformly for $-\varepsilon_0 \leq t \leq \varepsilon_0$ and η in a neighborhood of ξ_0. Here we write $g(x) = \mathcal{O}(\lambda^{-\infty})$ if $|g(\lambda)| \leq C_N \lambda^{-N}$ for every $N \geq 0$. On the other hand, 4) implies that

$$((D_t + P^*(x,D_x))(f\,e^{i\lambda\varphi})\,|\,u) = \mathcal{O}(\lambda^{-\infty}).$$

Hence,

$$F(0,w,\eta\,;\lambda) = F(t,w,\eta\,;\lambda) + \mathcal{O}(\lambda^{-\infty}).$$

If for some $t \in [-\varepsilon_0,\varepsilon_0]$ we have $\gamma(t) \notin WF(u)$, and if $\operatorname{supp} w \times \{\eta\}$ is sufficiently close to (x_0,ξ_0), then $F(t,w,\eta\,;\lambda) = \mathcal{O}(\lambda^{-\infty})$, by Proposition 8.2 and the geometric description of φ. Hence $F(0,w,\eta\,;\lambda) = \mathcal{O}(\lambda^{-\infty})$. Since $F(0,w,\eta\,;\lambda) = (w(x)\,e^{i\lambda x\eta}\,|\,u)_{L^2_x}$ if we let $w = 1$ near x_0, we see that $\gamma(0) \notin WF(u)$. □

If P is a differential operator, then we can avoid the use of pseudodifferential operators by working directly with the equation

$$\Big(\frac{1}{\lambda}D_t + \frac{1}{\lambda^m}P^*(x,D_x)\Big)(f\,e^{i\lambda\varphi}) = \mathcal{O}(\lambda^{-\infty}).$$

We end this chapter by studying the asymptotic expansions for pseudodifferential operators, used in the proof of Proposition 8.3. Let $P \in L^m_{1,0}(X)$ be properly supported with symbol $P(x,\xi)$, let $a \in S^k_{1,0}(X \times [1,\infty[)$ (in the sense that $|\partial_x^\alpha \partial_\lambda^j a(x\,;\lambda)| \leq C_{K,\alpha,j}\,\lambda^{k-j}$ for $x \in K \subset\subset X$, $\alpha \in \mathbb{N}^n$, $j \in \mathbb{N}$) and let $\varphi \in C^\infty(X\,;\mathbb{R})$ with $d\varphi \neq 0$ every where. We shall show that

$$I(x\,;\lambda) \stackrel{\mathrm{def}}{=} e^{-i\lambda\varphi(x)}\,P(a\,e^{i\lambda\varphi}) \in S^{m+k}_{1,0}(X \times [1,\infty[)$$

and give an asymptotic expansion for this quantity. Let $\chi_0 \in C^\infty(X \times X)$ be equal to 1 near the diagonal of $X \times X$ and have a proper support. Then modulo $S^{-\infty}(X \times [1,\infty[)$ (with the obvious definition)

$$I(x\,;\lambda) \equiv \frac{1}{(2\pi)^n} \iint e^{i(x-y)\eta + i\lambda(\varphi(y)-\varphi(x))}\,P(x,\eta)\,\chi_0(x,y)\,a(y\,;\lambda)\,dy\,d\eta.$$

If $x \in \tilde{X} \subset\subset X$ and $C > 1$ is sufficiently large, depending only on φ and $\tilde{X}$, then $I(x;\lambda)$ will be modified only by an element of class $S^{-\infty}(\tilde{X} \times [1,\infty[)$, if we introduce a cut-off function $\chi(\frac{\eta}{\lambda})$, where $\chi \in C_0^\infty(\dot{\mathbb{R}}^n)$ is equal to 1 on the shell $\frac{1}{C} \leq |\eta| \leq C$. (If suffices to integrate by parts in y.) After a change of variables we get modulo $S^{-\infty}(\tilde{X} \times [1,\infty[)$ for $x \in \tilde{X}$:

(8.1) $I(x;\lambda) \equiv$
$$\left(\frac{\lambda}{2\pi}\right)^n \iint e^{i\lambda((x-y)\eta + \varphi(y) - \varphi(x))} P(x,\lambda\eta)\chi(\eta)\chi_0(x,y)a(y,\lambda)\,dy\,d\eta.$$

We have the unique critical point $(y,\eta) = (x,\varphi'(x))$, which is non-degenerate, so the method of stationary phase and Proposition 1.9 show that $I(x;\lambda)$ belongs to $S_{1,0}^{m+k}(X \times [1,\infty[)$ and has an asymptotic expansion. To get an explicit expansion, we use the same trick as for the change of variables, and write $\varphi(x) - \varphi(y) = (x-y)\Phi(x,y)$, where $\Phi(x,y)$ is smooth in a neighborhood of supp χ_0 with values in $\mathbb{R}^n$, and satisfies $\Phi'(x,x) = \varphi'(x)$. After a change of variables in η, we get

$$I(x;\lambda) \equiv \left(\frac{\lambda}{2\pi}\right)^n \iint e^{-i\lambda(y-x)\eta} b(x,y,\eta;\lambda)\,dy\,d\eta \mod S^{-\infty}(\tilde{X} \times [1,\infty[),$$

where
$$b(x,y,\eta;\lambda) = P(x,\lambda(\Phi(y,x) + \eta))\chi(\eta + \Phi(y,x))\chi_0(x,y)a(y;\lambda)$$
$$\in S_{1,0}^{m+k}((\tilde{X} \times \tilde{X} \times \mathbb{R}^n) \times [1,\infty[)$$

has compact support in (y,η). Then as in Example 2.2, we get the asymptotic expansion in $S_{1,0}^{m+k}(X \times [1,\infty[)$

(8.2) $\displaystyle I(x;\lambda) \sim \sum_{\alpha \in \mathbb{N}^n} \frac{\lambda^{-|\alpha|}}{\alpha!} D_y^\alpha \partial_\eta^\alpha \left(P(x,\lambda(\Phi(y,x) + \eta))a(y,\lambda)\right)\Big|_{\substack{y=x \\ \eta=0}}.$

The term associated to α belongs to $S^{m+k-|\alpha|}$. For $\alpha = 0$, we get $P(x,\lambda\varphi'(x))a(x;\lambda)$. The sum of the terms with $|\alpha| = 1$ is

$$\sum_{j=1}^n \frac{1}{i\lambda} \partial_{y_j}\partial_{\eta_j}\left(P(x,\lambda(\Phi(y,x)+\eta))a(y,\lambda)\right)\Big|_{\substack{y=x \\ \eta=0}} =$$
$$\sum_{j=1}^n \frac{1}{i}P^{(j)}(x,\lambda\varphi'(x))\partial_{x_j}a(x;\lambda) + \sum_{j=1}^n\sum_{k=1}^n \frac{\lambda}{i}P^{(jk)}(x,\lambda\varphi'(x))\left(\frac{\partial \Phi_k}{\partial y_j}\right)_{y=x} a(x;\lambda),$$

where $P^{(j)} = \dfrac{\partial P}{\partial \eta_j}$, $P^{(jk)} = \dfrac{\partial^2 P}{\partial \eta_j \partial \eta_k}$. From $\varphi(y) - \varphi(x) = (y-x)\Phi(y,x)$, we get $\dfrac{\partial^2\varphi}{\partial x_j \partial x_k} = \left(\dfrac{\partial \Phi_k(y,x)}{\partial y_j} + \dfrac{\partial \Phi_j(y,x)}{\partial y_k}\right)\Big|_{y=x}$ so the last expression can be simplified and we get

$$I(x,\lambda) \equiv P(x,\lambda\varphi'(x))\,a(x,\lambda) + \Big(\sum_{j=1}^{n}\frac{1}{i}P^{(j)}(x,\lambda\varphi'_x(x))\partial_{x_j}$$
$$+\frac{\lambda}{2}\sum\sum\frac{1}{i}P^{(jk)}(x,\lambda\varphi'(x))\frac{\partial^2\varphi}{\partial x_j\partial x_k}\Big)a \bmod S_{1,0}^{m+k-2}(X\times\mathbb{R}_+).$$

This result can easily be extended to the case when a,φ,P depend smoothly on some parameter $\alpha\in\mathbb{R}^k$.

Remark : The function Φ is not unique but the asymptotic expansion (8.2) coincides with

(8.3) $$\qquad \sim \sum_{\beta}\frac{1}{\beta!}P^{(\beta)}(x,\lambda\varphi'(x))\,D_y^{\beta}(a(y\,;\lambda)\,e^{i\lambda\mathcal{H}(y,x)})\big|_{y=x}$$

where $\mathcal{H}(y,x) = \mathcal{O}(|y-x|^2)$ is defined by

$$\varphi(y) = \varphi(x) + (y-x)\,\varphi'(x) + \mathcal{H}(y,x).$$

$\Big($Here the general term is a symbol of order $\leq m+k-\frac{|\beta|}{2}.\Big)$ In fact, $\mathcal{H}(y,x) = (y-x)\cdot(\Phi(y,x)-\varphi'(x))$ so we can rewrite (8.3) as

$$\sim \sum\frac{1}{\beta!}P^{(\beta)}(x,\lambda\varphi'(x))D_y^{\beta}(e^{i\lambda(y-x)\cdot(\Phi(y,x)-\varphi'(x))}a(y;\lambda))\big|_{y=x}$$

$$= \sum\frac{1}{\beta!}P^{(\beta)}(x,\lambda\varphi'(x))(D_y-\lambda\varphi'(x))^{\beta}(e^{i\lambda(y-x)\cdot\Phi(y,x)}a(y;\lambda))\big|_{y=x}$$

$$= \sum\frac{1}{\beta!}P^{(\beta)}(x,\lambda\varphi'(x))\frac{1}{(2\pi)^n}\iint e^{i(x-y)(\eta-\lambda\Phi(y,x))}(\eta-\lambda\varphi'(x))^{\beta}a(y;\lambda)dyd\eta$$

$$= \sum\frac{1}{\beta!}P^{(\beta)}(x,\lambda\varphi'(x))\frac{1}{(2\pi)^n}\iint e^{i(x-y)\eta}(\eta+\lambda\Phi(y,x)-\lambda\varphi'(x))^{\beta}a(y;\lambda)dyd\eta$$

$$\sim \sum_{\beta}\frac{1}{\beta!}P^{(\beta)}(x,\lambda\varphi'(x))\sum_{\alpha}\frac{1}{\alpha!}D_y^{\alpha}\partial_{\eta}^{\alpha}((\eta+\lambda\Phi(y,x)-\lambda\varphi'(x))^{\beta}a(y;\lambda))\big|_{\substack{\eta=0\\y=x}}$$

$$= \sum_{\beta}\sum_{\alpha}\frac{1}{\alpha!}D_y^{\alpha}\partial_{\eta}^{\alpha}\Big(\frac{1}{\beta!}P^{(\beta)}(x,\lambda\varphi'(x))(\eta+\lambda\Phi(y,x)-\lambda\varphi'(x))^{\beta}a(y;\lambda)\Big)\big|_{\substack{\eta=0\\y=x}}$$

(iterated asymptotic sum). Here we notice that the general term, depending on both α and β, is of order $\leq m+k-|\alpha|$, and vanishes for $2|\alpha|<|\beta|$. Hence the general term is of order $\leq m+k-\max\big(|\alpha|,\frac{|\beta|}{2}\big)$, and it is then clear that we can change the order of summation in the last expression. The β-sum is just a Taylor sum and we see that (8.3) reduces to (8.2).

Exercises

Exercise 8.1

Let $u \in \mathcal{D}'(\mathbb{R}^4)$. Assume u is C^∞ in a neighborhood of the axis $x_1 = x_2 = x_3 = 0$. Assume $\partial_{x_1} u \in C^\infty(\mathbb{R}^4)$ and $(\partial_{x_2} - x_1 \partial_{x_3}) u \in C^\infty(\mathbb{R}^4)$. Show that $u \in C^\infty(\mathbb{R}^4)$.

Exercise 8.2

Let P be the differential operator on $\mathbb{R}^2$:

$$P = (x_2 D_{x_2})^2 - D_{x_1}^2 + f(x) D_{x_1} + g(x) D_{x_2} + h(x)$$

where $f, g, h \in C^\infty(\mathbb{R}^2)$.

1) Determine the bicharacteristic strip of P passing through $(x_1, x_2; \xi_1, \xi_2) = (0, 1; 1, 1)$ up to reparametrization. (Note that the principal symbol can be factorized.)

2) Determine all the bicharacteristics of P passing above $x = (0, 1)$ and the projections on $\mathbb{R}_x^2$.

3) Show that if Pu is C^∞ in $\Omega = \{(x_1, x_2); x_2 > 0\}$ and u is C^∞ in $\{(x_1, x_2); 0 < x_2 < \varepsilon, |x_1| > \varepsilon^{-1}\}$ for some $\varepsilon > 0$ then u is C^∞ in a neighborhood of $x = (0, 1)$.

4) Consider the particular case

$$P = (x_2 D_{x_2})^2 - D_{x_1}^2.$$

Writing P as a product of operators of order 1, show that if $Pu \in C^\infty(\mathbb{R}^2)$ and u is C^∞ in a neighborhood of $(0, 0)$ then u is C^∞ in a neighborhood of $\{(x_1, x_2); x_2 = 0\}$.

Exercise 8.3

On $\mathbb{R}^2$, let $v(x_1, x_2)$ be defined by

$$v(x_1, x_2) = 1 \quad \text{if} \quad x_1 > \frac{1}{2} x_2^2$$

$$v(x_1, x_2) = 0 \quad \text{if} \quad x_1 \leq \frac{1}{2} x_2^2$$

1) Check that $(\partial_{x_2} + x_2 \partial_{x_1}) v = 0$.

2) Determine $WF(v)$ and show that this set is a Lagrangian submanifold of $T^*\mathbb{R}^2$.

3) Let $v_k = (\partial_{x_2} - x_2 \partial_{x_1})^k v$, $k \in \mathbb{N}$.
Show that $WF(v_k) = WF(v)$.

4) Let $p(x,\xi) = -\xi_2^2 + x_2^2 \xi_1^2 \in C^\infty(T^*\mathbb{R}^2)$ and let H_p be the Hamiltonian vector field of p.
Calculate H_p and the integral curve γ of H_p passing through $\rho = (0,1;1,1)$. What is the projection of γ on $\mathbb{R}^2_x$?

5) Let $P_a(x,D) = \partial_{x_2}^2 - x_2^2 \partial_{x_1}^2 + a \partial_{x_1}$, $a \in \mathbb{R}$. Let $u \in \mathcal{D}'(\mathbb{R}^2)$ satisfy $P_a u \in C^\infty(\mathbb{R}^2)$. Assume that u is C^∞ in a neighborhood of the line $x_2 = 0$. Show that $u \in C^\infty(\mathbb{R}^2)$.

6) Show that $P_1 v = 0$ and that $P_{2k+1} v_k = 0$.

Exercise 8.4

Let $p(x,\xi) = \xi_2^2 - x_2 \xi_1^2 \in C^\infty(T^*\mathbb{R}^2 \setminus 0)$, $x = (x_1, x_2) \in \mathbb{R}^2$, $\xi = (\xi_1, \xi_2) \in \mathbb{R}^2$ and let H_p be the Hamiltonian vector field of p.

1) Determine the integral curve γ of H_p passing through $\rho = (0,0;1,0)$.

2) In a neighborhood of ρ, define Λ as the C^∞ conic (in ξ) 2-dimensional submanifold of $T^*\mathbb{R}^2$ containing γ.
Show that Λ can be parametrized near ρ by

$$x_1 = -\frac{\partial H(\xi)}{\partial \xi_1} \quad x_2 = -\frac{\partial H(\xi)}{\partial \xi_2} \quad \text{where} \quad H(\xi) = -\frac{1}{3}\frac{\xi_2^3}{\xi_1^2}.$$

Show that p vanishes on Λ and that Λ is Lagrangian.

3) Let $a(x,\xi) \in S^0_{1,0}(\mathbb{R}^2 \times \mathbb{R}^2)$ be supported in a small conic neighborhood of ρ in $\mathbb{R}^2 \times \dot{\mathbb{R}}^2$. Show that the oscillatory integral $I = \int e^{i(x\xi + H(\xi))} a(x,\xi) d\xi$ is well defined in $\mathcal{D}'(\mathbb{R}^2)$.
Determine $WF(I)$.
Show how to find a such that $\rho \notin WF(P(I))$, where $P = P(x,D) = D_{x_2}^2 - x_2 D_{x_1}^2$.

4) Let $u \in \mathcal{D}'(\mathbb{R}^2)$ and assume that $Pu \in C^\infty(\mathbb{R}^2)$. Assume that u is C^∞ in $x_1 < 0$. Show that u is C^∞ in $\mathbb{R}^2$.

Notes

Theorem 8.1 is due to Hörmander. Many results on propagation of singularities have subsequently been obtained in more and more complicated situations, as well as in the framework of analytic and Gevrey singularities. See [Hö4].

9 Local symplectic geometry II

Lemma 9.1 Let f, g be C^2 functions, defined on some open subset of T^*X, where X is a smooth manifold. Then $H_{\{f,g\}} = [H_f, H_g]$ (where $[H_f, H_g] = H_f H_g - H_g H_f$ when H_f, H_g are viewed as first-order differential operators).

Proof: We need to prove that $[H_f, H_g] \lrcorner \sigma = -d\{f, g\}$. On one hand $\mathcal{L}_{H_f}(H_g \lrcorner \sigma) = -\mathcal{L}_{H_f} dg = -d\{f, g\}$ and on the other hand

$$\mathcal{L}_{H_f}(H_g \lrcorner \sigma) = [H_f, H_g] \lrcorner \sigma + H_g \lrcorner \mathcal{L}_{H_f}(\sigma) = [H_f, H_g] \lrcorner \sigma,$$

so we get the desired identity. $\square$

Lemma 9.2 (The Jacobi identity) If f, g, h are C^2 functions, then $\{f, \{g, h\}\} + \{g, \{h, f\}\} + \{h, \{f, g\}\} = 0$.

Proof: Using the antisymmetry of the Poisson bracket and Lemma 9.1, we get

$$-\{f, \{g, h\}\} = \{\{g, h\}, f\} = H_{\{g,h\}} f$$
$$= [H_g, H_h] f = \{g, \{h, f\}\} - \{h, \{g, f\}\}$$
$$= \{g, \{h, f\}\} + \{h, \{f, g\}\}.$$

$\square$

By definition, a symplectic manifold is a manifold (which is sufficiently smooth and in our case C^∞) carrying a smooth closed differential 2-form σ which is non-degenerate in the sense that it induces a non-degenerate alternate bilinear form on the tangent space at each point of the manifold. We call σ the *symplectic form*. The cotangent bundle of a manifold is an obvious example of such a manifold. A symplectic manifold is necessarily of even dimension $= 2n$, and all our definitions and results so far remain valid with T^*X replaced by a more general symplectic manifold. The (smooth) local coordinates $q_1, \ldots, q_n, p_1, \ldots, p_n$ are called *symplectic* or sometimes *canonical* if $\sigma = \sum_{1}^{n} dp_j \wedge dq_j$.

Proposition 9.3 The local coordinates (q, p) are symplectic if and only if

(9.1) $\{q_j, q_k\} = 0, \ \{p_j, p_k\} = 0, \ \{p_j, q_k\} = \delta_{j,k}, \ \forall j, k \in \{1, \ldots, n\}$.

Proof: If (q,p) are symplectic coordinates, we obtain as in the special case of canonical coordinates on a cotangent bundle, that for smooth functions f, g :

(9.2) $$\{f, g\} = \sum_1^n \frac{\partial f}{\partial p_j} \frac{\partial g}{\partial q_j} - \frac{\partial f}{\partial q_j} \frac{\partial g}{\partial p_j},$$

and (9.1) follows. Conversely, if the local coordinates (q,p) satisfy (9.1), we obtain (9.2) by Taylor expanding f, g to the first order, and hence for every smooth function f

(9.3) $$H_f = \sum \frac{\partial f}{\partial p_j} \frac{\partial}{\partial q_j} - \frac{\partial f}{\partial q_j} \frac{\partial}{\partial p_j}.$$

Using the identity $\sigma(H_f, H_g) = \{f, g\}$ it then follows that $\sigma = \sum_1^n dp_j \wedge dq_j$. $\square$

The Darboux theorem implies the existence of local symplectic coordinates :

Theorem 9.4 *Let (M, σ) be a $2n$-dimensional symplectic manifold. Let $J, K \subset \{1, \ldots, n\}$ and let p_j, $j \in J$, q_k, $k \in K$, be real-valued smooth functions with linearly independent differentials, defined in a neighborhood of $p_0 \in M$, satisfying the three equations of (9.1) (for all j, k for which the brackets are defined). Then we can find smooth functions p_j, $j \in \{1, \ldots, n\} \setminus J$, q_k, $k \in \{1, \ldots, n\} \setminus K$, defined near p_0, such that $q_1, \ldots, q_n, p_1, \ldots, p_n$ are local symplectic coordinates.*

Proof: Since the Hamilton field of a constant vanishes, it follows from Lemma 9.1 that the vector fields H_{p_j}, H_{q_k} commute with each other and we can then find local coordinates $(z, x, y) = (z_1, \ldots, z_{2n-\#J-\#K}, (x_j)_{j \in J}, (y_k)_{k \in K})$ centered at p_0 (i.e. vanishing at p_0) such that $H_{p_j} = \frac{\partial}{\partial x_j}$, $H_{q_k} = \frac{\partial}{\partial y_k}$. In fact, it suffices to choose a submanifold F of dimension $2n - \#J - \#K$ passing through p_0 transversally to the space spanned by the H_{p_j}, H_{q_k} at that point, and impose that the restrictions of x_j, y_k to F vanish and that the restrictions of the z_ℓ form a system of local coordinates.

Let F be given by $x = 0$, $y = 0$ and assume for instance that $n \notin K$. Then (in view of (9.1)) we are looking for $q_n = q_n(z, x, y)$ such that

$$\begin{cases} \dfrac{\partial}{\partial x_j} q_n = \delta_{j,n}, & j \in J \\[6pt] \dfrac{\partial}{\partial y_k} q_n = 0, & k \in K \end{cases}$$

This system has a unique solution (near ρ_0) if we prescribe $q_n|_F$. If $n \in J$ we can take any such q_n, and dp_j, dq_k, $j \in J$, $k \in K \cup \{n\}$ are still linearly independent, as can be seen by applying $\dfrac{\partial}{\partial x_n}$ to a linear combination. Iterating this argument we can extend the set of functions q_j, p_k until $J = K$. In the case $J = K$ we assume again for instance that $n \notin K$. We then define F by $p_j = q_j = 0$, $j \in J$ and check that F is transverse to the space spanned by H_{p_j}, H_{q_j}, $j \in J$. It then suffices to take q_n such that $H_{p_j} q_n = H_{q_j} q_n = 0$, $j \in J$ with $dq_n|_F \neq 0$. The last condition implies that dq_n, dp_j, dq_j, $j \in J$ are independent. □

As already mentioned, the preceding theorem with $K = J = \emptyset$ guarantees the existence of local symplectic coordinates so a symplectic manifold can always be locally identified with $T^*\mathbb{R}^n$, but even in the case of $T^*\mathbb{R}^n$ one is frequently interested in other symplectic coordinates than the standard ones.

For the applications to partial differential equations, when $M = T^*X \backslash 0$, one also frequently wants the symplectic coordinates to be *homogeneous* in the sense that

$$p_j(x, \lambda\xi) = \lambda p_j(x,\xi), \quad q_j(x,\lambda\xi) = q_j(x,\xi), \quad \lambda > 0.$$

(Other types of homogeneity assumption sometimes appear, and a more general setting is to consider homogeneous symplectic manifolds (M, σ, m_λ), where m_λ, $\lambda > 0$ are diffeomorphisms $M \to M$ depending smoothly on λ such that $m_{\lambda\mu} = m_\lambda \circ m_\mu$, $m_\lambda^* \sigma = \lambda\sigma$.)

Let $J, K \subset \{1, \ldots, n-1\}$ and let p_j, q_k, $j \in J$, $k \in K$ be smooth functions defined in a conic neighborhood of a point $\rho_0 = (x_0, \xi_0)$, $\xi_0 \neq 0$, with linearly independent differentials, satisfying the homogeneity assumption above as well as (9.1) (for all the Poisson brackets that are defined), and assume also that $p_j(\rho_0) = q_k(\rho_0) = 0$, $j \in J$, $k \in K$.

Then if H_{p_j}, H_{q_k}, $\xi \cdot \dfrac{\partial}{\partial \xi}$ for $j \in J\backslash(J \cap K)$, $k \in K\backslash(J \cap K)$ are linearly independent, it is possible to complete the p_j, q_k into a system of homogeneous symplectic coordinates, p_j, q_k, $j, k \in \{1, \ldots, n\}$, with $p_j(\rho_0) = 0$, $1 \leq j \leq n-1$, $p_n(\rho_0) = 1$, $q_k(\rho_0) = 0$, $1 \leq k \leq n$.

Proof: We follow the proof of Theorem 9.4. Let Σ be a conic manifold defined by $p_j = q_j = 0$ for $j \in J \cap K$. One can verify that Σ is a symplectic manifold equipped with the 2-form $\sigma|_\Sigma$, or equivalently that $T\Sigma \pitchfork T\Sigma^\perp$ where $\pitchfork =$ "is transversal to", that H_{p_j}, H_{q_k}, $j, k \in J \cap K$ span $T\Sigma^\perp$ at every point of Σ and that H_{p_j}, H_{q_k} are tangent to Σ for $j \in J\backslash(J \cap K)$, $k \in K\backslash(J \cap K)$. Since $\xi \dfrac{\partial}{\partial \xi} \in T\Sigma$ the hypothesis of linear independence implies that $\xi \cdot \dfrac{\partial}{\partial \xi}$,

H_{p_j}, $j \in J$, H_{q_k}, $k \in K$, are independent. Choosing F conic, we can then complete our system as in the first case of the proof of Theorem 9.4, provided that we prescribe homogeneous functions on F. In this way we can complete until $J = K$. Suppose for instance that $J = K = \{1, \ldots, d\}$, $d \leq n-1$. Then $\Sigma = \{p_j = q_j = 0\}$ is symplectic, conic of dimension $2n - 2d$. If $d < n-1$ we can find $\tilde{p}_{d+1}$ on Σ, homogeneous of degree 1, and vanishing at ρ_0, such that $H^\Sigma_{\tilde{p}_{d+1}}$ (the Hamilton field on Σ for $\sigma|_\Sigma$) is independent from $\xi \cdot \dfrac{\partial}{\partial \xi}$. This is possible since we have $2n - 2d - 1 > 1$ degrees of freedom in the choice of $d\tilde{p}_{d+1}$ at ρ_0. After that we construct p_{d+1} solving all the relevant equations in (9.1) and with $p_{d+1}|_\Sigma = \tilde{p}_{d+1}$. Then $H_{p_{d+1}} = H^\Sigma_{\tilde{p}_{d+1}}$ on Σ so the hypothesis of linear independence will be satisfied.

If $d = n-1$, we take $\tilde{p}_{d+1}(\rho_0) = 1$. Then $H^\Sigma_{\tilde{p}_{d+1}}$ and $\xi \cdot \dfrac{\partial}{\partial \xi}$ are independent and we conclude as before. (Notice that in this case we cannot take $\tilde{p}_{d+1}(\rho_0) = 0$, because Σ is now of dimension 2, and then $\xi \cdot \dfrac{\partial}{\partial \xi}$ and $H^\Sigma_{\tilde{p}_{d+1}}$ would be parallel at ρ_0.) $\square$

If (M, σ_M) and (N, σ_N) are symplectic manifolds of the same dimension and if $\mathcal{H} : M \to N$ is a C^∞ map, we call $\mathcal{H}$ a canonical transformation if $\mathcal{H}^* \sigma_N = \sigma_M$. A canonical transformation is necessarily a local diffeomorphism for the following reason. If $2n = \dim M$, then the n-fold wedge product $\sigma_M \wedge \ldots \wedge \sigma_M = \sigma_M^n$ is a non-vanishing differential form on M of maximal degree 2^n, which in local symplectic coordinates can be written

$$\sigma_M^n = n! \, dp_1 \wedge dq_1 \wedge dp_2 \wedge dq_2 \wedge \ldots \wedge dp_n \wedge dq_n.$$

If $\mathcal{H}$ is canonical, then $\mathcal{H}^*(\sigma_N^n) = \sigma_M^n$ and hence the Jacobian of $\mathcal{H}$ is 1 if we choose local symplectic coordinates in M and in N.

It is easy to check that the following alternative characterizations of canonical transformations hold :

- $\mathcal{H}$ is a canonical transformation if and only if $\{f \circ \mathcal{H}, g \circ \mathcal{H}\} = \{f, g\} \circ \mathcal{H}$ for all C^∞ functions on N. (Here the first bracket is the Poisson bracket on M while the second is that of N.)

- In the case $N = T^* \mathbb{R}^n$ (or an open subset of $T^* \mathbb{R}^n$), then $\mathcal{H}$ is a canonical transformation if and only if $q_j = x_j \circ \mathcal{H}$, $p_j = \xi_j \circ \mathcal{H}$ form a system of local symplectic coordinates (near any point of M).

- $\mathcal{H}$ is a canonical transformation if and only its graph $C_\mathcal{H} = \{(\mathcal{H}(\rho), \rho) \, ; \, \rho \in M\} \subset N \times M$ is a Lagrangian submanifold, when $N \times M$ is considered as a symplectic manifold equipped with the 2-form $\sigma_N - \sigma_M$. (Strictly speaking we should denote the latter 2-form by $i^* \sigma_N - j^* \sigma_M$, where $i : N \times M \to N$, $j : N \times M \to M$ are the canonical projections.)

The second characterization gives a link between the problem of constructing canonical transformations and the Darboux theorem. Using the third characterization, we are naturally led to the useful notion of generating functions for canonical transformations. Indeed, let N, M be open sets in $T^*\mathbb{R}^n$ and $\mathcal{H} : M \ni (y,\eta) \mapsto (x,\xi) \in N$ a canonical transformation, so that the graph $C_\mathcal{H}$ is a Lagrangian manifold for $\sum d\xi_j \wedge dx_j - \sum d\eta_j \wedge dy_j = \sum d\xi_j \wedge dx_j + \sum dy_j \wedge d\eta_j$. If near some point of $C_\mathcal{H}$, the map $C_\mathcal{H} \ni (x,\xi,y,\eta) \mapsto (x,\eta) \in \mathbb{R}^{2n}$ is a local diffeomorphism, then (by Theorem 5.3) $C_\mathcal{H}$ is given near that point by the equations $\xi = \dfrac{\partial \varphi}{\partial x}(x,\eta)$, $y = \dfrac{\partial \varphi}{\partial \eta}(x,\eta)$ for some smooth function $\varphi(x,\eta)$. In other words, locally $\mathcal{H}$ is of the form $\left(\dfrac{\partial \varphi}{\partial \eta}(x,\eta),\eta\right) \mapsto \left(x,\dfrac{\partial \varphi}{\partial x}(x,\eta)\right)$. The fact that $C_\mathcal{H}$ is (locally) the graph of a smooth map is then equivalent to the fact that $\det \dfrac{\partial^2 \varphi}{\partial x\, \partial \eta} \neq 0$. We call φ a (local) generating function for $\mathcal{H}$. We shall end this chapter by showing that if $(x_0,\xi_0,y_0,\eta_0) \in C_\mathcal{H}$ and $\eta_0 \neq 0$, then possibly after a change of y-coordinates (and the corresponding dual change of η-coordinates), $\mathcal{H}$ has a generating function near that point. We will start with some preparations in the linearized situation.

By definition a *symplectic space* is a real vector space of finite dimension $2n$ which is equipped with an alternate bilinear form which is non-degenerate. A typical example of a symplectic space is the tangent space at some point of a symplectic manifold. We also define in the obvious way Lagrangian (sub)spaces and linear canonical transformations between symplectic spaces. There is a linear Darboux theorem (which can be proved by linear algebra by imitating the proof of Theorem 9.4), which tells us that a general symplectic space (M,σ_M) can be identified by means of a linear canonical transformation with $(\mathbb{R}^n \times \mathbb{R}^{n*},\sigma)$, where $\sigma((x,\xi),(y,\eta)) = \langle \xi, y \rangle - \langle \eta, x \rangle$. (It is here convenient to distinguish between $\mathbb{R}^n$ and its dual $\mathbb{R}^{n*}$, in order to indicate that we may actually replace $\mathbb{R}^n$ by a general n-dimensional real vector space E.)

If $\Lambda \subset \mathbb{R}^n \times \mathbb{R}^{n*}$ is a Lagrangian space transversal to $\{(x,\xi) \in \mathbb{R}^n \times \mathbb{R}^{n*} \,;\, x = 0\}$ (which is also a Lagrangian space), then Λ is of the form $\Lambda_A : \xi = Ax$ where A is a real symmetric matrix. Conversely every such Λ_A is Lagrangian. (We may further notice that $Ax = \dfrac{\partial}{\partial x}\left(\dfrac{1}{2}\langle Ax, x\rangle\right)$, and that everything we said remains valid and meaningful with $\mathbb{R}^n$ replaced by E. A is then symmetric $E \to E^*$.)

If $\Lambda_0 \subset \mathbb{R}^n \times \mathbb{R}^{n*}$ is a Lagrangian space, then to say that $\Lambda_A : \xi = Ax$ and Λ_0 are not transversal is to a polynomial condition on the coefficients of A. In fact, let Λ_0 be of the form $K(x,\xi) = 0$, where $K : \mathbb{R}^n \times \mathbb{R}^{n*} \to \mathbb{R}^n$ is surjective. Then, to say that $\Lambda_0 \cap \Lambda_A \neq \{0\}$ is to say that the determinant of the map $\mathbb{R}^n \ni x \mapsto K(x, Ax)$ is equal to zero.

With Λ_0 as above, there exists at least one Λ_A which is transverse to Λ_0 :

Lemma 9.5 *There exists an invertible $n \times n$ matrix H such that after the canonical change of coordinates $(x, \xi) \mapsto (Hx, {}^tH^{-1}\xi)$, Λ_0 takes the form $x' = 0$, $\xi'' = Bx''$, where B is a symmetric matrix. Here we write $x = (x', x'')$, $\xi = (\xi', \xi'')$, $x' = (x_1, \ldots, x_{n-d})$, $x'' = (x_{n-d+1}, \ldots, x_n)$, for some suitable d with $0 \leq d \leq n$.*

Proof: Let $L \subset \mathbb{R}^n_x$ be the projection of Λ_0. After a change of coordinates in x (and the corresponding dual change in ξ), we may assume that L is given by $x' = 0$. Then Λ_0 contains no non-vanishing vector of the form $v_0 = (x'_0, 0, 0, \xi''_0)$. In fact, if v_0 is such a vector, we first see that $x'_0 = 0$, and since $\sigma(v_0, (x, \xi)) = \langle \xi''_0, x'' \rangle = 0$ for all $(x, \xi) \in \Lambda_0$, we next see that $\xi''_0 = 0$, since x'' may be arbitrary in $\mathbb{R}^d$.

The map $\Lambda_0 \ni (x, \xi) \mapsto (x'', \xi') \in \mathbb{R}^n$ is then bijective and Λ_0 is of the form $x' = 0$, $\xi'' = Bx'' + C\xi'$. Finally we have $C = 0$, because if we consider $((0, 0), (\xi'_0, C\xi'_0)) \in \Lambda_0$, we know that for all $(x, \xi) \in \Lambda_0$:

$$0 = \sigma((0, (\xi'_0, C\xi'_0)), (x, \xi)) = \langle C\xi'_0, x'' \rangle = 0,$$

and moreover B is symmetric. □

With $Ax = (0, Dx'')$, we see that Λ_A is transversal to Λ_0 if $\det(D - B) \neq 0$ (assuming that Λ_0 has been brought to the form in the lemma), and it is then clear that there exists at least one symmetric matrix A such that Λ_A is transversal to Λ_0, and the polynomial in A given by $\det(x \mapsto K(x, Ax))$, that we found prior to Lemma 9.4, is therefore not identically equal to zero. (Another way of proving that this polynomial is not identically zero, is to use complex Lagrangian spaces in $\mathbb{C}^n \times \mathbb{C}^{n*}$. See Exercise 9.7)

Let now $\Lambda \subset T^*X \backslash 0$ be a Langrangian manifold and $(x_0, d\varphi(x_0)) \in \Lambda$, $(d\varphi(x_0) \neq 0)$, where X is a smooth n-dimensional manifold. If we choose local coordinates x near x_0 and the corresponding dual coordinates ξ, then we can consider the smooth map $\pi_\xi : (x, \xi) \mapsto \xi$ from a neighborhood of $(x_0, d\varphi(x_0))$ in Λ to $\mathbb{R}^n$. Contrary to the case of the base projection $T^*X \backslash 0 \to X$, the map π_ξ is not invariantly defined, and we are interested in choosing the canonical coordinates (x, ξ) in such a way that π_ξ becomes a local diffeomorphism. For a choice of x-coordinates above, π_ξ is a local diffeomorphism at $(x_0, d\varphi(x_0)) = (x_0, \xi_0)$ iff the Lagrangian manifold, given by $\xi = \xi_0$, intersects Λ transversally at (x_0, ξ_0).

Choose the x-coordinates centered at x_0 such that $\xi_0 = (0, \ldots, 0, 1)$. (After fixing a choice of φ which a priori is defined only up to $\mathcal{O}(x^2)$, we may choose $x_1, \ldots, x_n$ so that $x_n = \varphi(x)$.) Then $T_{(x_0, \xi_0)}(T^*X)$ can be naturally

identified with $\mathbb{R}^n \times \mathbb{R}^{n*}$ and $\Lambda_0 = T_{(x_0,\xi_0)}(\Lambda)$ becomes a Lagrangian subspace of $\mathbb{R}^n \times \mathbb{R}^{n*}$. According to Lemma 9.5 and the adjacent discussion, we can then find a symmetric matrix A, such that $\Lambda_A = \{\xi = Ax\}$ is transversal to Λ_0. Now Λ_A is the tangent space at (x_0, ξ_0) of $\Lambda_{\tilde{\varphi}} = \{\xi = \tilde{\varphi}'(x)\}$, where $\tilde{\varphi}(x) = x_n + \frac{1}{2}\langle Ax, x\rangle$ (with $d\tilde{\varphi}(x_0) = d\varphi(x_0)$), so $\Lambda_{\tilde{\varphi}}$ and Λ intersect transversally at $(x_0, d\varphi(x_0))$. Now choose new local coordinates $(y_1, \ldots, y_n)$ centered at x_0 with $y_n = \tilde{\varphi}$. Then according to the previous remarks, if $(y_1, \ldots, y_n; \eta_1, \ldots, \eta_n)$ are the canonical coordinates associated to $(y_1, \ldots, y_n)$, then the map $\pi_\eta : (y, \eta) \mapsto \eta$ is a local diffeomorphism from a neighborhood of $(x_0, d\varphi(x_0))$ to $\mathbb{R}^n$. Then (cf. Theorem 5.3), there is a smooth function $H(\eta)$ defined near $\eta = (0, \ldots, 0, 1)$ such that Λ is given by
$$-y = \frac{\partial H}{\partial \eta}(\eta)$$
in (the (y, η)-coordinates in) a neighborhood of $(x_0, d\varphi(x_0))$.

Let now X and Y be two smooth manifolds of the same dimension n and let $\mathcal{H}$ be a canonical transformation from a neighborhood of a point $(y_0, \eta_0) \in T^*Y \backslash 0$ onto a neighborhood of some point $(x_0, \xi_0) = \mathcal{H}(y_0, \eta_0)$ in T^*X. The fiber $T^*_{x_0}X$ over the point x_0 is a Lagrangian manifold and hence $\Lambda = \mathcal{H}^{-1}(T^*_{x_0}X)$ has the same property. Since $\eta_0 \neq 0$ we obtain with a convenient choice of local coordinates $y_1, \ldots, y_n$ near y_0, that the projection $\pi_\eta : \Lambda \to \mathbb{R}^n_\eta$ is a local diffeomorphism. The projection $\pi_{x,\eta}$: graph$(\mathcal{H}) \cap$ neighborhood of $(x_0, \xi_0; y_0, \eta_0) \ni (x, \xi; y, \eta) \mapsto (x, \eta) \in X \times \mathbb{R}^n$ is then also a local diffeomorphism and it follows that $\mathcal{H}$ can be described by a generating function $\varphi(x, \eta)$.

Remark : If $\omega = \sum \xi_j \, dx_j$ is the canonical 1-form on T^*X, then $\langle \omega, t\rangle = \sigma\left(\xi \cdot \frac{\partial}{\partial \xi}, t\right)$ for all $t \in T(T^*X \backslash 0)$. (Here $\xi \cdot \frac{\partial}{\partial \xi}$ is an invariantly defined vector field on T^*X.) If Λ is a *conic* Lagrangian manifold (in the sense that $(x, \xi) \in \Lambda \Longrightarrow (x, \lambda\xi) \in \Lambda$, for every $\lambda > 0$) then $\omega|_\Lambda = 0$. If the local coordinates $x_1, \ldots, x_n$ are chosen so that $\pi_\xi : \Lambda \to \mathbb{R}^n_\xi$ is a local diffeomorphism near some fixed point (x_0, ξ_0), then with $H(\xi) = -x\xi|_\Lambda$ (with $x = x(\xi)$ describing Λ) we get $dH = -\xi \, dx|_\Lambda - x \, d\xi|_\Lambda = -x \, d\xi|_\Lambda$ so in other words $-x = \frac{\partial H}{\partial \xi}$ on Λ. Similarly if $\mathcal{H}$ is a homogeneous canonical transformation, in the sense that $(x, \xi) = \mathcal{H}(y, \eta) \Longrightarrow (x, \lambda\xi) = \mathcal{H}(y, \lambda\xi)$ for every $\lambda > 0$, and if the y-coordinates are choosen so that $\pi_{x,\eta}$ is a local diffeomorphism from a neighborhood of some point in graph$(\mathcal{H})$ to $X \times \mathbb{R}^n$, then a generating function for $\mathcal{H}$ is given by $\varphi(x, \eta) = y\eta|_{\text{graph}(\mathcal{H})}$.

Exercises

Exercise 9.1

Let u, v be vector fields. Show that $\mathcal{L}_v u = [v, u]$. Deduce that

$$(\exp tv)_*(u) = u + \int_0^t (\exp sv)_*([v, u])\, ds.$$

Hint for the first part : we may assume that $v \neq 0$ at the point of interest.

Exercise 9.2

a) Let v_1, v_2 be two vector fields defined in a neighborhood of $0 \in \mathbb{R}^n$. Suppose that $[v_1, v_2] = a_1(x)\, v_1 + a_2(x)\, v_2$, $a_1, a_2 \in C^\infty$ (Frobenius condition).
If $v_1(0) \neq 0$, $v_2(0) \neq 0$ show that there exist $b_1, b_2 \in C^\infty$ with $b_1(0) \neq 0$, $b_2(0) \neq 0$, such that $[b_1 v_1, b_2 v_2] = 0$ in a neighborhood of 0.

b) Let $v_1, \ldots, v_k$ be vector fields defined in a neighborhood of 0 in $\mathbb{R}^n$, satisfying $[v_i, v_j] = 0$, $\forall i, j$. Show that for $t \in \mathbb{R}^k$, $x \in \mathbb{R}^n$ small enough we have

$$\exp(t_1 v_1) \circ \ldots \circ \exp(t_k v_k)(x) = \exp(t_1 v_1 + \ldots + t_k v_k)(x).$$

If furthermore $v_1(0), \ldots, v_k(0)$ are linearly independent, show that there exist local coordinates near 0 such that $v_j = \dfrac{\partial}{\partial x_j}$, $1 \leq j \leq k$.

Exercise 9.3

What is the degree of homogeneity of the symplectic form $\sum_1^n d\xi_j \wedge dx_j$?

a) for the dilatation group $(x, \xi) \to (x, \lambda \xi)$, $\lambda \in \mathbb{R}^+$.

b) for the dilatation group $(x, \xi) \to (\lambda^{1/2} x, \lambda^{1/2} \xi)$, $\lambda \in \mathbb{R}^+$.

Exercise 9.4

a) Let A be a complex $n \times n$ matrix with spectrum in $\{z \in \mathbb{C};\ \mathrm{Re}\, z > 0\}$. Show that $\mathbb{C}^n$ can be equipped with an inner product $(x \mid y)$ such that $\mathrm{Re}(Ax \mid x) \geq \dfrac{1}{C} \|x\|^2$, $x \in \mathbb{C}^n$, for some positive constant C. Here $\|x\| = \sqrt{(x \mid x)}$.

b) Deduce that $\|\exp(-tA)(x)\| \leq e^{-t/C} \|x\|$, $x \in \mathbb{C}^n$, $t \geq 0$.

c) Let $v = \sum v_j(x)\, \partial_{x_j}$ be a real smooth vector field defined in a neighborhood of a point $x_0 \in \mathbb{R}^n$ and vanishing at that point. Define the linearized vector field by $v_0 = \sum_1^n \sum_1^n \dfrac{\partial v_j(x_0)}{\partial x_k} x_k\, \partial_{x_j}$, and the corresponding matrix $F = \left(\dfrac{\partial v_j(x_0)}{\partial x_k}\right)$. Show that F can be viewed as a well defined endomorphism $T_{x_0}(\mathbb{R}^n) \to T_{x_0}(\mathbb{R}^n)$ independent of the choice of local coordinates.

d) With v, F as in c), assume that the spectrum of F is contained in $\{z \in \mathbb{C}; \operatorname{Re} z > 0\}$. Show that there is a Hilbert space norm $\|\cdot\|$ on $\mathbb{R}^n$ and a constant $C > 0$ such that $\|x - x_0\| \leq \dfrac{1}{C} \implies \exp(-tv)(x)$ is well defined for $t \geq 0$, and moreover

$$\|\exp(-tv)(x) - x_0\| \leq e^{-t/C}\|x - x_0\|, \quad t \geq 0.$$

e) Study the differential of $x \mapsto \exp(-tv)(x)$, when $t \to +\infty$.

f) What can be said about F when v is a Hamilton field?

Exercise 9.5

Let $\Omega \subset \mathbb{R}^n$ be an open subset and v a vector field defined on Ω, vanishing at $x_0 \in \Omega$ and such that the linearized vector field of v at x_0 has its spectrum in $\{\operatorname{Re} z > 0\}$. Assume furthermore that, for every $x \in \Omega$, $\exp(-tv)(x)$ is well defined and belongs to Ω for every $t \geq 0$, that $\exp(-tv)(x) \to x_0$, $t \to \infty$, and that $\{\exp(-tv)(x) ; x \in K, t \geq 0\}$ is relatively compact in Ω for every compact $K \subset \Omega$. (Find an example.) Let ω be a k-form ($1 \leq k \leq n$) C^∞ on Ω with $d\omega = 0$. Show that $\eta = \displaystyle\int_0^\infty (\exp(-tv))^*(v \lrcorner \omega)\, dt$ is a $(k-1)$-form C^∞ on Ω with $d\eta = \omega$.

Exercise 9.6

Let $v = \sum a_j(x)\partial_{x_j}$ be a vector field on an open subset Ω of $\mathbb{R}^n$. Calculate $\mathcal{L}_v(dx_1 \wedge \ldots \wedge dx_k)$ and show that for every compact subset $K \subset \Omega$ there exists $t_K > 0$ such that

$$(\exp tv)^*(dx_1 \wedge \ldots \wedge dx_n) = dx_1 \wedge \ldots \wedge dx_n \quad \text{on} \quad K, \quad \text{for} \quad |t| < t_K,$$

iff $\operatorname{div}(v) \stackrel{\text{def}}{=} \displaystyle\sum_1^n \dfrac{\partial a_j}{\partial x_j}(x) = 0$ on Ω.

Exercise 9.7

If L is a linear subspace of $\mathbb{R}^n_x \times \mathbb{R}^n_\xi$ we let $L^\mathbb{C} = L + iL = \{u + iv ; u, v \in L\} \subset \mathbb{C}^n_z \times \mathbb{C}^n_\zeta$ denote its complexification. Here we write $z = x + iy$, $\zeta = \xi + i\eta$. A complex-linear subspace $\mathcal{L} \subset \mathbb{C}^n \times \mathbb{C}^n$ is called Lagrangian if $\dim_\mathbb{C} \mathcal{L} = n$ (where $\dim_\mathbb{C}$ denotes the complex dimension) and if $\sigma^\mathbb{C}|_\mathcal{L} = 0$ where $\sigma^\mathbb{C} = \displaystyle\sum_1^n d\zeta_j \wedge dz_j$ is the complex symplectic form. Let $\Lambda_0 \subset \mathbb{R}^n_x \times \mathbb{R}^n_\xi$ be a Lagrangian subspace.

a) Show that $\Lambda_0^\mathbb{C}$ is a complex Lagrangian subspace.

b) A complex Lagrangian subspace $\Lambda \subset \mathbb{C}^n \times \mathbb{C}^n$ is called strictly positive if $\dfrac{1}{i}\langle \sigma^\mathbb{C}, t \wedge \bar{t}\rangle > 0$ for every $0 \neq t \in \Lambda$. Show that $\Lambda_A : \zeta = Az$, $z \in \mathbb{C}^n$ defines a strictly positive Lagrangian subspace if A is a symmetric $n \times n$ matrix with $\operatorname{Im} A > 0$ (in the sense of self-adjoint matrices, where $\operatorname{Im} A = \dfrac{1}{2i}(A - \bar{A})$).

c) Show that if $\Lambda \subset \mathbb{C}^n \times \mathbb{C}^n$ is a strictly positive Lagrangian subspace, then $\Lambda \cap \Lambda_0^{\mathbb{C}} = \{0\}$.

(Hint : compute $\dfrac{1}{i} \langle \sigma^{\mathbb{C}}, t \wedge \bar{t} \rangle$ for $t \in \Lambda_0^{\mathbb{C}}$.)

d) Deduce the converse statement in b) by choosing $\Lambda_0 : x = 0$.

e) Show, without making use of Lemma 9.5, that there is a real symmetric $n \times n$ matrix A such that $\Lambda_0 \cap \Lambda_A = \{0\}$.

Notes

Same as for Chapter 5, except possibly for Exercise 9.7, which evokes some basic notions in the theory of Fourier integral operators with complex phase functions. See [Ma2], [Hö4,7], [MS].

10 Canonical transformations of pseudodifferential operators

Let $\mathcal{H}$ be a homogeneous canonical transformation from a conic neighborhood of $(y_0, \eta_0) \in T^*\mathbb{R}^n \setminus 0$ onto a conic neighborhood of $(x_0, \xi_0) \in T^*\mathbb{R}^n \setminus 0$, with $\mathcal{H}(y_0, \eta_0) = (x_0, \xi_0)$. Possibly after a change of local coordinates in the y-variables, we may assume that $\mathcal{H}$ is given by the homogeneous generating function $\varphi(x, \eta)$, defined in a conic neighborhood of (x_0, η_0). Let $a \in S^0_{1,0}(\mathbb{R}^{2n}_{x,y} \times \mathbb{R}^n_\eta)$ be supported in a sufficiently small conic neighborhood of (x_0, y_0, η_0) in $\mathbb{R}^{2n}_{x,y} \times \overset{n}{\mathbb{R}}_\eta$. We can then introduce the Fourier integral operator

(10.1) $$Au(x) = \iint e^{i(\varphi(x,\eta) - y\eta)} a(x, y, \eta) u(y) \, dy \, d\eta,$$

and we shall consider that A is associated to $\mathcal{H}$. This is justified by the fact that $WF'(A) \subset \text{graph}(\mathcal{H})$, and the aim of this chapter is to give an even stronger justification. Similarly, if $b \in S^0_{1,0}(\mathbb{R}^{2n}_{y,x} \times \mathbb{R}^n_\eta)$ is supported in a small conic neighborhood of (y_0, x_0, η_0), we consider that the Fourier integral operator

(10.2) $$Bv(y) = \iint e^{i(y\eta - \varphi(x,\eta))} b(y, x, \eta) v(x) \, dx \, d\eta$$

is associated to $\mathcal{H}^{-1}$. (One possible choice of B would be $B = A^*$.) Then $WF'(B) \subset \text{graph}(\mathcal{H}^{-1})$.

For $u \in C_0^\infty(\mathbb{R}^n)$, we find

(10.3) $$ABu(x) = \iiiint e^{i(\varphi(x,\zeta) - z\zeta + z\theta - \varphi(y,\theta))} a(x, z, \zeta) b(z, y, \theta) u(y) \, dz \, d\zeta \, dy \, d\theta$$

where the integral can be viewed as the limit in $\mathcal{D}'$, when we approach a, b by sequences of functions in $S^{-\infty}$. Using integrations by parts in z, we see that we can insert a cutoff $\chi\left(\dfrac{(\zeta - \theta)}{|\theta|}\right)$ in the integral in (10.3) without changing the operator mod $L^{-\infty}$. Using the method of stationary phase for the z, ζ-integrations, where we have the non-degenerate critical point given by $\zeta = \theta$, $z = \varphi'_\zeta(x, \zeta)$, we get

(10.4) $$AB\,u(x) = \iint e^{i(\varphi(x,\theta) - \varphi(y,\theta))} c(x, y, \theta) u(y) \, dy \, d\theta + Ku,$$

with $K \in L^{-\infty}$, where $c \in S^0_{1,0}$ is supported in a small conic neighborhood of (x_0, x_0, η_0). In a conic neighborhood of $\text{supp}\,c$, we can write

$$\varphi(x, \theta) - \varphi(y, \theta) = (x - y) \Xi(x, y, \theta)$$

where $\theta \mapsto \Xi(x, y, \theta)$ is C^∞ (also in x, y), homogeneous of degree 1 and with an inverse having the same properties. (Here we apply the implicit function theorem and the fact that $\Xi(x, x, \theta) = \varphi'_x(x, \theta)$ so that $\dfrac{\partial}{\partial \theta} \Xi(x, x, \theta) =$

$\varphi_{x\theta}''(x,\theta)$ is invertible, and assume that a and b have sufficiently small support so that c is supported in a sufficiently small conic neighborhood of (x_0, x_0, η_0).) After a change of variables, we get

$$AB\,u(x) = \iint e^{i(x-y)\xi}\,\tilde{c}(x,y,\xi)\,u(y)\,dy\,\frac{d\xi}{(2\pi)^n} + Ku,$$

where $\tilde{c} \in S_{1,0}^0$ has its support in a small conic neighborhood of (x_0, x_0, ξ_0).

Hence $AB \in L_{1,0}^0(\mathbb{R}^n)$ and moreover, if a, b are non-characteristic at (x_0, y_0, η_0) and (y_0, x_0, ξ_0) respectively, then AB is non-characteristic at (x_0, ξ_0).

Similarly, we can show that $BA \in L_{1,0}^0$ and if a, b are non-characteristic as above, then BA is non-characteristic at (y_0, η_0).

With a, b non-characteristic as above we know that AB is non-characteristic at (x_0, ξ_0), and hence according to Chapter 4 we can find $R \in L_{1,0}^0(\mathbb{R}^n)$, properly supported, such that the symbol of ABR is equal to 1 mod $S^{-\infty}$, in a conic neighborhood of (x_0, ξ_0), or equivalently $(x_0, \xi_0) \notin WF(ABR - I)$. We express this by saying that $ABR \equiv I$ near (x_0, ξ_0). If $\tilde{B} = BR$, then using the method of stationary phase as at the end of Chapter 8, we obtain that

$$\tilde{B}\,u(y) = \iint e^{-i(y\eta - \varphi(x,\eta))}\,\tilde{b}(y,x,\eta)\,u(x)\,dx\,d\eta + Ku,$$

with $K \in L^{-\infty}$ and $\tilde{b} \in S_{1,0}^0$. Moreover

$$A\tilde{B} \equiv I \quad \text{near} \quad (x_0, \xi_0).$$

Similarly, we can construct $\hat{B}$ with the same phase as $\tilde{B}$ and with a symbol $\hat{b} \in S_{1,0}^0$, such that

$$\hat{B}A \equiv I \quad \text{near} \quad (y_0, \eta_0).$$

Then $(y_0, \eta_0, x_0, \xi_0) \notin WF'(\tilde{B} - \hat{B})$. In fact, modulo operators C with $(y_0, \eta_0, x_0, \xi_0) \notin WF'(C) \subset \text{graph}(\mathcal{H}^{-1})$, we get

$$\hat{B} \equiv \hat{B}A\tilde{B} \equiv \tilde{B}.$$

It follows that

$$\tilde{B}A \equiv I \quad \text{near} \quad (y_0, \eta_0).$$

We have then established a considerable part of the Egorov theorem :

Theorem 10.1 *Let $\mathcal{H}$ be a homogeneous canonical transformation from a conic neighborhood of $(y_0, \eta_0) \in T^*\mathbb{R}^n \backslash 0$ onto a conic neighborhood of $(x_0, \xi_0) \in T^*\mathbb{R}^n \backslash 0$, with $\mathcal{H}(y_0, \eta_0) = (x_0, \xi_0)$, given by a generating function $\varphi(x, \eta)$ (positively homogeneous of degree 1 in η). Let $A : \mathcal{D}'(\mathbb{R}^n) \to \mathcal{D}'(\mathbb{R}^n)$,*

$C^\infty(\mathbb{R}^n) \to C^\infty(\mathbb{R}^n)$, be of the form (10.1) with $a \in S_{1,0}^0$ non-characteristic at (x_0, y_0, η_0) and with support in a sufficiently small conic neighborhood of that point. Then we can find $B : \mathcal{D}'(\mathbb{R}^n) \to \mathcal{D}'(\mathbb{R}^n)$, $C^\infty(\mathbb{R}^n) \to C^\infty(\mathbb{R}^n)$ of the form (10.2) with $b \in S_{1,0}^0$ non-characteristic at (y_0, x_0, η_0) and with support in a small conic neighborhood of that point, such that BA, $AB \in L_{1,0}^0(\mathbb{R}^n)$ and $BA \equiv I$ near (y_0, η_0), $AB \equiv I$ near (x_0, ξ_0). Moreover, if $P \in L_{1,0}^m(\mathbb{R}^n)$, then $Q = BPA \in L_{1,0}^m(\mathbb{R}^n)$, $(x_0, \xi_0, y_0, \eta_0) \notin WF'(PA - AQ)$ and for the principal symbols we have the relation $q = p \circ \mathcal{H}$ in a conic neighborhood of (y_0, η_0).

End of the proof : It remains to verify the last statement in the theorem. To see that $Q \in L_{1,0}^m$ it suffices to notice that $PA = \tilde{A}$ is of the form (10.1) with a replaced by $\tilde{a}(x, y, \eta) \equiv p(x, \varphi_x'(x, \eta))\, a(x, y, \eta)$ mod $S_{1,0}^{m-1}$. As already indicated, we compute

$$BAu(z) = \iiiint e^{i(z\zeta - \varphi(x,\zeta) + \varphi(x,\eta) - y\eta)} b(z, x, \zeta)\, a(x, y, \eta)\, u(y)\, dy\, d\eta\, dx\, d\zeta$$
$$= \iint e^{i(z-y)\eta} c(z, y, \eta)\, u(y)\, dy\, \frac{d\eta}{(2\pi)^n} + Ku(z),$$

$K \in L^{-\infty}$, by the method of stationary phase (for the (x, ζ)-integrations). The corresponding critical point is given by $\zeta = \eta$, $\varphi_\eta'(x, \eta) = z$ and

$$c(y, y, \eta) \equiv f_{b,\varphi}(y, \eta)\, a(x, y, \eta)|_{\varphi_\eta'(x, \eta) = y} \quad \text{mod } S_{1,0}^{-1},$$

where $f_{b,\varphi} \in S_{1,0}^0$. If we replace A by $\tilde{A}$, then c is replaced by $\tilde{c}$, with

$$\tilde{c}(y, y, \eta) \equiv f_{b,\varphi}(y, \eta)\, (p(x, \varphi_x'(x, \eta))\, a(x, y, \eta))|_{\varphi_\eta'(x, \eta) = y}$$
$$\equiv p(x, \varphi_x'(x, \eta))|_{\varphi_\eta'(x, \eta) = y} \cdot c(y, y, \eta) \quad \text{mod } S_{1,0}^{m-1}.$$

Here $c(y, y, \eta)$ mod $S_{1,0}^{-1}$ is the principal symbol of BA and $\tilde{c}(y, y, \eta)$ mod $S_{1,0}^{m-1}$ that of $B\tilde{A}$. In a conic neighborhood of (y_0, η_0) we have $c(y, y, \eta) \equiv 1$ mod $S_{1,0}^{-1}$ and hence

$$\tilde{c}(y, y, \eta) \equiv p(x, \varphi_x'(x, \eta))|_{\varphi_\eta'(x, \eta) = y} = (p \circ \mathcal{H})(y, \eta) \quad \text{mod } S_{1,0}^{m-1}.$$

The verification of "$(x_0, \xi_0, y_0, \eta_0) \notin WF'(PA - AQ)$" is left as an exercise. $\square$

It is an instructive exercise to try to understand directly how P and Q should be related in order to have $(x_0, \xi_0, y_0, \eta_0) \notin WF'(PA - AQ)$. We notice that the theorem above remains valid when all operators have classical symbols.

Remark 10.2 If f is a diffeomorphism from a neighborhood of a point $z_0 \in \mathbb{R}^n$ onto a neighborhood of a point $y_0 \in \mathbb{R}^n$, then the corresponding pull-back operation :

$$f^*u = u \circ f(z) = \iint e^{i(f(z)-y)\cdot\eta} u(y)\, dy\, \frac{d\eta}{(2\pi)^n}$$

can be viewed as a Fourier integral operator of the form (10.1) associated to the canonical transformation

$$\mathcal{H}_{f^{-1}} : (f(x), \eta) \mapsto (x, {}^t\!f'(x)\eta),$$

given by the generating function $f(x)\cdot\eta$, and the result on changes of variables in Chapter 3 can therefore be viewed (essentially) as a special case of Egorov's theorem.

Let $\mathcal{H}$ be a canonical transformation as in Theorem 10.1, but drop the assumption about the existence of a generating function. As noticed at the end of Chapter 9, we can then (possibly after shrinking the neighborhoods of (y_0, η_0) and (x_0, ξ_0) between which $\mathcal{H}$ is a diffeomorphism), write $\mathcal{H} = \mathcal{H}_1 \circ \mathcal{H}_2$, where $\mathcal{H}_1$ has a generating function and where $\mathcal{H}_2 = \mathcal{H}_{f^{-1}}$ for a suitable local diffeomorphism f as above.

To $\mathcal{H}_1$ we associate an operator A_1 as in the theorem and to $\mathcal{H}_2$ we associate $A_2 = f^*$. Let B_1 be given by the theorem and let $B_2 = A_2^{-1}$. Then with $A = A_1 \circ A_2$, $B = B_2 \circ B_1$ the conclusions of the theorem remain valid and moreover

$$WF'(A) \subset \text{graph}(\mathcal{H}), \quad WF'(B) \subset \text{graph}(\mathcal{H}^{-1}).$$

Example 1 : Let $P \in L^1_{\text{cl}}(\mathbb{R}^n)$ have the real principal symbol $p(x, \xi)$ (positively homogeneous of degree 1 in ξ). Let $(x_0, \xi_0) \in T^*\mathbb{R}^n \backslash 0$ be a point where $p(x_0, \xi_0) = 0$ and where $dp(x_0, \xi_0)$ and $\Sigma\, \xi_j\, dx_j$ are independent. (The last assumption can be reformulated as : "$H_p(x_0, \xi_0)$ and $\xi \cdot \dfrac{\partial}{\partial \xi}$ are independent".) Then according to the homogeneous version of the Darboux theorem we can find a homogeneous canonical transformation from a conic neighborhood of (y_0, η_0) onto a conic neighborhood of (x_0, ξ_0) such that $p \circ \mathcal{H} = \eta_n$. Here $(y_0, \eta_0) = (0, (1, 0, \ldots, 0))$. Let $Q \in L^1_{\text{cl}}$ be an operator as in Theorem 10.1 (extended as in Remark 10.2). Modifying the symbol of Q outside a conic neighborhood of (y_0, η_0), we find $R \in L^0_{\text{cl}}(\mathbb{R}^n)$ such that

$$PA \equiv A(D_{x_n} + R) \quad \text{near} \quad (x_0, \xi_0\,;\, y_0, \eta_0),$$

in the sense that $(x_0, \xi_0\,;\, y_0, \eta_0) \in WF'(PA - A(D_{x_n} + R))$. (Logically we should write $D_{y_n} + R(y, D_y)$ instead of $D_{x_n} + R(x, D_x)$ but we switch to the more standard notations.)

We next try to improve A in order to eliminate R.

Lemma 10.3 There exists an elliptic operator $C \in L^0_{cl}(\mathbb{R}^n)$, such that $(D_{x_n} + R)C \equiv C D_{x_n}$.

Proof: We need to solve $[D_{x_n}, C] + RC \equiv 0 \mod L^{-\infty}$. If we introduce the full symbols $\sigma_C \sim c_0 + c_{-1} + \ldots$, $\sigma_R \sim r_0 + r_{-1} + \ldots$ where c_{-j}, r_{-j} are positively homogeneous of degree $-j$ (in ξ), our equation becomes equivalent to a sequence of transport equations :

$$\frac{1}{i}\frac{\partial c_0}{\partial x_n} + r_0 c_0 = 0$$

$$\frac{1}{i}\frac{\partial c_{-1}}{\partial x_n} + r_0 c_{-1} = f_1(c_0)$$

$$\vdots$$

If we impose the initial conditions $c_0 = 1$, $c_{-1} = c_{-2} = \ldots = 0$ on the hyperplane $x_n = 0$, we get a unique solution of this system with c_0 non-vanishing. □

Let $D \in L^0_{cl}(\mathbb{R}^n)$ be a parametrix of C. Putting $\tilde{A} = AC$, $\tilde{B} = DB$, we get

$$P\tilde{A} \equiv \tilde{A} D_{x_n} \quad \text{near} \quad (x_0, \xi_0\,;\,y_0, \eta_0)$$
$$\tilde{B}\tilde{A} \equiv I \quad \text{near} \quad (y_0, \eta_0)$$
$$\tilde{A}\tilde{B} \equiv I \quad \text{near} \quad (x_0, \xi_0).$$

In this situation, we can say that "P is microlocally equivalent to D_{x_n}". Using this result one can re-prove the result of Chapter 8 on propagation of singularities (exercise), and we shall use such a method for another class of operators in Example 2. Let us show that P is microlocally surjective :

There exists a conic neighborhood V of (x_0, ξ_0) such that for every $v \in \mathcal{D}'(\mathbb{R}^n)$, there exists $u \in \mathcal{D}'(\mathbb{R}^n)$ with $V \cap WF(Pu - v) = \emptyset$.

Proof: Let $v \in \mathcal{D}'(\mathbb{R}^n)$ and put $\tilde{v} = \tilde{B}v \in \mathcal{E}'(\mathbb{R}^n)$. If $H(x_n)$ denotes the Heaviside function (i.e. the characteristic function for the positive half-axis), then the convolution

$$\tilde{u} = (\delta(x') \otimes H(x_n)) * \tilde{v} = \tilde{E}\tilde{v},$$

solves the equation $D_{x_n} \tilde{u} = \tilde{v}$. In general if $w \in \mathcal{D}'(\mathbb{R}^n)$ then for the corresponding operator of convolution with w we have

$$WF'(w*) \subset \{(x, \xi\,;\,y, \eta)\,;\,\xi = \eta,\ (x - y, \eta) \in WF(w)\},$$

since the distribution kernel of $w*$ is f^*w (formally $= w(x-y)$), where $f : (x,y) \mapsto x - y$. In our case, $w = \delta(x') \otimes H(x_n)$ and
$$WF(w) \subset \{(x,\xi);\ x = 0\} \cup \{(x,\xi);\ \xi_n = 0,\ x' = 0,\ x_n > 0\}.$$
Hence,
$$WF'(\tilde{E}) \subset \text{diag}((T^*\mathbb{R}^n\setminus 0) \times (T^*\mathbb{R}^n\setminus 0)) \cup \\ \{((x', x_n + t, \xi', 0);\ (x, \xi', 0));\ t \geq 0\},$$
so
$$WF(\tilde{u}) \subset WF(\tilde{v}) \cup \{(x', x_n + t, \xi', 0); (x, \xi', 0) \in WF(\tilde{v}),\ t \geq 0\}.$$
We put $u = \tilde{A}\tilde{u} = Ev$ with $E = \tilde{A}\tilde{E}\tilde{B}$. Notice first that
$$WF'(E) \subset \text{diag}((T^*\mathbb{R}^n\setminus 0) \times (T^*\mathbb{R}^n\setminus 0)) \cup \\ \{(\exp t H_p(\rho), \rho);\ p(\rho) = 0,\ t \geq 0\},$$
since $WF'(E) \subset \mathcal{H} \circ WF'(\tilde{E}) \circ \mathcal{H}^{-1}$ and $\exp(t H_{\xi_n})(x, \xi', 0) = (x', x_n+t, \xi', 0)$, $\mathcal{H} \circ \exp t H_{\xi_n} \circ \mathcal{H}^{-1} = \exp t H_p$. It follows that
$$WF(u) \subset WF(v) \cup \{\exp t H_p(\rho);\ p(\rho) = 0,\ \rho \in WF(v),\ t \geq 0\}.$$

Let us finally check that $Pu \equiv v$ in V (i.e. $V \cap WF(Pu - v) = \emptyset$) if V is sufficiently small. Since $D_{x_n} \circ \tilde{E} = I$, we can write
$$PE = (P\tilde{A} - \tilde{A} D_{x_n})\tilde{E}\tilde{B} + \tilde{A}\tilde{B}.$$
Using that $(x_0, \xi_0;\ y_0, \eta_0) \notin WF'(P\tilde{A} - \tilde{A}D_{x_n}) \subset \text{graph}(\mathcal{H})$, it is easy to see that if V is sufficiently small, then $WF((P\tilde{A} - \tilde{A}D_{x_n})w) \cap V = \emptyset$ for every $w \in \mathcal{D}'$. Moreover $\tilde{A}\tilde{B} \equiv I$ and hence
$$Pu = (P\tilde{A} - \tilde{A}D_{x_n})\tilde{E}\tilde{B}v + \tilde{A}\tilde{B}v \equiv 0 + v \quad \text{in} \quad V.$$

$\square$

We call E a microlocal parametrix. There also exists a second microlocal parametrix F with
$$WF'(F) \subset \text{diag}((T^*\mathbb{R}^n\setminus 0) \times (T^*\mathbb{R}^n\setminus 0)) \cup \\ \{(\exp t H_p(\rho), \rho);\ p(\rho) = 0,\ t \leq 0\}.$$
The theorem on propagation of singularities implies that E and F are (essentially) unique. Under suitable assumptions on the behavior of the bicharacteristic strips of P, it is possible (using pseudodifferential partitions of unity) to construct local and sometimes even global parametrices.

Example 2: Let $P \in L^1_{cl}(\mathbb{R}^n)$ have the homogeneous principal symbol $p(x,\xi)$ vanishing at a point $(x_0,\xi_0) \in T^*\mathbb{R}^n\backslash 0$ and assume that in a conic neighborhood V of (x_0,ξ_0) (to which we restrict our attention from now on)

(10.5) $p(x,\xi) = 0 \implies d\operatorname{Re} p, d\operatorname{Im} p, \xi \cdot dx$ are linearly independent.

Then $\Sigma \stackrel{\text{def}}{=} \{(x,\xi) \in V\,;\, p(x,\xi) = 0\}$ is a conic submanifold of codimension 2. The Poisson bracket

$$\frac{1}{i}\{p,\bar{p}\} = -2\{\operatorname{Re} p, \operatorname{Im} p\}$$

will now play an important role. In the present example, we shall assume that

(10.6) $\dfrac{1}{i}\{p,\bar{p}\} = 0$ on Σ.

Then Σ is an *involutive* manifold, where we pause to recall the definition and some properties of such manifolds: a submanifold $J \subset T^*X\backslash 0$ is said to be involutive if $T_\rho J^\perp \subset T_\rho J$ for every $\rho \in J$. Equivalently if we represent Γ locally by the (real) equations $f_1(x,\xi) = \ldots = f_d(x,\xi) = 0$ (where $f_1,\ldots,f_d$ are C^∞ functions with $df_1,\ldots,df_d$ independent at every point), then $\{f_j,f_k\} = 0$ on J. In fact, $T_\rho J^\perp$ is generated by $H_{f_1},\ldots,H_{f_d}$ for every $\rho \in J$, and the requirement that the Poisson bracket vanishes on J is equivalent to the requirement that $H_{f_1},\ldots,H_{f_d}$ are tangent to J. We now assume J involutive. Then in a neighborhood of any given point in J we get by Taylor's formula (after completing $f_1,\ldots,f_d$ into a system of local coordinates)

$$\{f_j,f_k\} = \sum_{\nu=1}^d a^\nu_{j,k}(x,\xi) f_\nu, \text{where}\quad a^\nu_{j,k} \in C^\infty.$$

Then *on J*:

$$[H_{f_j},H_{f_k}] = \sum_{\nu=1}^d a^\nu_{j,k}(x,\xi) H_{f_\nu}.$$

Viewing H_{f_j} as vector fields on J, we recognize here the Frobenius integrability condition. Applying the Frobenius theorem (cf. Exercise 9.2 in the case when $d = 2$), we can find near any given point of J local coordinates (s,t) in J, with $s \in \mathbb{R}^d$, $t \in \mathbb{R}^{2n-2d}$ centered at that point, such that the submanifolds $\Gamma_{t_0} : t = t_0$ are integral manifolds of $H_{f_1},\ldots,H_{f_d}$, in other words $T_\rho \Gamma_{t_0} = T_\rho J^\perp$ for every $\rho \in \Gamma_{t_0}$. We call Γ_{t_0} a bicharacteristic leaf.

Returning to our example, we know that Σ is involutive of codimension 2, so that the corresponding bicharacteristic leaves are 2-dimensional (and nowhere tangent to $\xi \cdot \partial_\xi$). Possibly after decreasing V, we may assume that the bicharacteristic leaves are closed connected submanifolds of Σ.

Theorem 10.4 *Under the assumptions above, let $\Gamma \subset \Sigma$ be a bicharacteristic leaf and let $u \in \mathcal{D}'(\mathbb{R}^n)$ satisfy $WF(Pu) \cap \Gamma = \emptyset$. Then either $\Gamma \cap WF(u) = \emptyset$ or $\Gamma \subset WF(u)$.*

We shall prove the theorem by transforming P microlocally into $D_{x_{n-1}} + iD_{x_n}$ and then by proving the theorem for this particular operator. We may assume that $(x_0, \xi_0) \in \Gamma$ and we will work in a small conic neighborhood of that point.

Lemma 10.5 *There exists a smooth non-vanishing function $a(x, \xi)$, homogeneous of degree 0, defined in a conic neighborhood of (x_0, ξ_0), such that $\{\bar{a} p, a \bar{p}\} = 0$.*

Proof:

$$\{\bar{a} p, a \bar{p}\} = \bar{a} a \{p, \bar{p}\} + \bar{a} \bar{p} \{p, a\} + a p \{\bar{a}, \bar{p}\} + p \bar{p} \{\bar{a}, a\}.$$

Since $\{p, \bar{p}\}$ is imaginary and vanishes on Σ we can find $b \in C^\infty$ homogeneous of degree 0, such that

$$\{p, \bar{p}\} = \bar{b} p - b \bar{p}.$$

Hence,

$$\begin{aligned}\{\bar{a} p, a \bar{p}\} &= \bar{a} \bar{p}\Big(\{p, a\} - ab + \frac{1}{2}\frac{p}{\bar{a}}\{\bar{a}, a\}\Big) \\ &\quad - ap\Big(\{\bar{p}, \bar{a}\} - \bar{a}\bar{b} + \frac{1}{2}\frac{\bar{p}}{a}\{a, \bar{a}\}\Big)\end{aligned}$$

and it suffices to find a non-vanishing smooth function a, homogeneous of degree 0 such that

(10.7) $$H_p a - b a + \frac{1}{2}\frac{p}{\bar{a}}\{\bar{a}, a\} = 0.$$

Restricting this equation to Σ we get

(10.8) $$H_p a - b a = 0 \quad \text{on} \quad \Sigma,$$

and this equation can further be restricted to any of the 2-dimensional characteristic leaves and it then becomes an *elliptic* partial differential equation of order 1. We shall then use the following fact.

Let $L = a_1(x)\dfrac{\partial}{\partial x_1} + a_2(x)\dfrac{\partial}{\partial x_2} + b(x)$ be an elliptic differential operator with C^∞ coefficients on $\overline{\Omega}$, where $\Omega \subset \mathbb{R}^2$ (for instance) is the open unit disk. Then, for every $v \in C^\infty(\overline{\Omega})$, there exists $u \in C^\infty(\overline{\Omega})$ such that $Lu = v$. Moreover, if $L = L_t$, $v = v_t$ depend smoothly on some parameters $t = (t_1, \ldots, t_d) \in$ some open set in $\mathbb{R}^d$, then we can choose the solution $u = u_t$ with the same property. (This result can be proved by establishing first an a priori inequality, $\|u\|_{H^1} \leq C\|Lu\|_{H^0}$, for $u \in C_0^\infty(\Omega)$, with the help of which we then show that the Dirichlet problem for L^*L is well posed, in the sense that $L^*L : H_0^1(\Omega) \to H^{-1}(\Omega)$ is bijective. We refrain from developing the details.)

On a bicharacteristic leaf, diffeomorphic to the unit disk, we can now solve the equation $H_p f = b$, and $a = e^f$ will then be a non-vanishing solution of (10.8) along that leaf. If we let $H \subset \Sigma$ be a hypersurface transversal to $\xi \cdot \partial_\xi$ which is also a union of bicharacteristic leaves, then we still have a smooth solution f of $H_p f = b$ on H, and we can extend f to a full conic neighborhood of (x_0, ξ_0) by requiring f to be homogeneous of degree 0. Then $a_0 = e^f$ will be a solution of (10.8). (Our arguments are local so we may have to shrink V and consequently Σ, but there is a choice of V such that (10.8) and (10.10) below will have smooth solutions for any smooth data $c_{\nu\mu}$.)

It follows that $\{\bar{a}_0 p, a_0 \bar{p}\}$ vanishes to the second order on Σ, and after replacing p by $\bar{a}_0 p$, we may assume that $\{p, \bar{p}\}$ vanishes to the second order on Σ. Assume by induction that we have found a smooth function a_k homogeneous of degree 1 with $a_k|_\Sigma = 1$ such that $H_p a_k - b a_k + \frac{1}{2} \frac{p}{\bar{a}_k} \{\bar{a}_k, a_k\}$ vanishes to order $k+1$ on Σ. Then we can write

(10.9) $$H_p a_k - b a_k + \frac{1}{2} \frac{p}{\bar{a}_k} \{\bar{a}_k, a_k\} = \sum_{\nu+\mu=k+1} c_{\nu\mu} p^\nu \bar{p}^\mu,$$

where $c_{\nu\mu}$ are smooth and homogeneous. We try

$$a_{k+1} = a_k - \sum_{\nu+\mu=k+1} d_{\nu\mu} p^\nu \bar{p}^\mu.$$

Then (10.9) improves by one step in k if we choose $d_{\nu\mu}$ in such a way that

(10.10) $$H_p d_{\nu\mu} - b d_{\nu\mu} = c_{\nu\mu} \quad \text{on} \quad \Sigma.$$

Again this equation can be solved and we get (10.9) for $k = 0, 1, 2, \ldots$. Also notice that $a_{k+\ell} - a_k$ vanishes to order $k+1$ on Σ for every $\ell \geq 1$.

By a Borel procedure we can find $\tilde{a} \in C^\infty(V)$, homogeneous of degree 0, such that $\tilde{a} - a_k$ vanishes to order $k+1$ on Σ for every k, and we then know that $\tilde{a}|_\Sigma = 1$ and that $\{\bar{\tilde{a}} p, \tilde{a} \bar{p}\}$ vanishes to infinite order on Σ. After replacing p by $\bar{\tilde{a}} p$, we may then assume that $\{p, \bar{p}\}$ vanishes to infinite order on Σ. Put $h = \frac{1}{i} \{p, \bar{p}\}$. Then

$$H_{\operatorname{Re} p}(\operatorname{Im} p) = -\frac{h}{2},$$

and we can find a real-valued function g (homogeneous etc.) vanishing to infinite order on Σ such that $H_{\operatorname{Re} p} g = \frac{h}{2}$. If $\hat{p} = \operatorname{Re} p + i(\operatorname{Im} p + g)$, then

$\{\hat{p}, \overline{\hat{p}}\} = 0$. On the other hand, $\hat{p} = \left(\frac{\hat{p}}{p}\right)p$, where $\frac{\hat{p}}{p}$ is smooth in V since $\hat{p} - p$ vanishes to infinite order on Σ. □

Let a be as in Lemma 10.5. Applying the Darboux theorem in its homogeneous version we can find a homogeneous canonical transformation $\mathcal{H}$ from a conic neighborhood of $(y_0, \eta_0) = (0, (1, 0, \ldots, 0))$ onto a conic neighborhood of (x_0, ξ_0), such that $(ap) \circ \mathcal{H} = \xi_{n-1} + i\,\xi_n$.

Let $A \in L^0_{cl}(\mathbb{R}^n)$ have principal symbol $= a$ in a conic neighborhood of (x_0, ξ_0). Let F be an operator associated to $\mathcal{H}$ as in Theorem 10.1 and Remark 10.2 with the microlocal inverse G, associated to $\mathcal{H}^{-1}$. Then

$$APF \equiv F(D_{x_{n-1}} + i\,D_{x_n} + R)$$

microlocally near $(x_0, \xi_0;\, y_0, \eta_0)$. Here $R \in L^0_{cl}(\mathbb{R}^n)$.

As in Example 1, we can conjugate $(D_{x_{n-1}} + i\,D_{x_n} + R)$ with a suitable elliptic pseudodifferential operator in order to eliminate R. (The transport equations will now be of Cauchy-Riemann type.) Hence, after modifying F and G, we may assume that $R = 0$.

Let u, Γ be as in the theorem. After restricting Γ to a sufficiently small conic neighborhood of (x_0, ξ_0) (independent of u), we have

$$WF((D_{x_{n-1}} + i\,D_{x_n})Gu) \cap \mathcal{H}^{-1}(\Gamma) = \emptyset,$$

and for $(y, \eta) \in \Gamma$ we have $(y, \eta) \in WF(Gu)$ if and only if $\mathcal{H}(y, \eta) \in WF(u)$. It then suffices to prove the theorem with P replaced by $(D_{x_{n-1}} + i\,D_{x_n})$ and with Γ replaced by $\mathcal{H}^{-1}(\Gamma)$. Without loss of generality we may assume that Γ passes through $(0, \eta_0)$ and then $\mathcal{H}^{-1}(\Gamma)$ will be of the form $\xi_1 = 1$, $\xi_2 = \ldots = \xi_n = 0$, $x_1 = \ldots = x_{n-2} = 0$, $(x_{n-1}, x_n) \in W$, where W is an open neighborhood of 0.

Let $V_a \subset T^*\mathbb{R}^n \setminus 0$ be the conic neighborhood of $(0, \eta_0)$, given by $(x_1, x_2) \in W$, $|x_j| < a$, $\xi_1 > 0$, $\left|\frac{\xi_k}{\xi_1}\right| < a$, $3 \le j \le n$, $2 \le k \le n$. Let $u \in \mathcal{D}'(\mathbb{R}^n)$ with $V_a \cap WF((D_{x_{n-1}} + i\,D_{x_n})u) = \emptyset$. Let $\delta << a$ and let $\chi \in C^\infty_0(\mathbb{R}^n)$ have its support in $|x| < \delta$, and be equal to 1 near 0. We write $z = y_{n-1} + i\,y_n$ and we put

$$\Phi_{z,\xi}(x) := (\chi(\cdot)\,e^{-i(\cdot)\xi})(x_1, \ldots, x_{n-2}, x_{n-1} - y_{n-1}, x_n - y_n).$$

Then

$$\begin{aligned}
2\frac{\partial}{\partial \bar{z}}\langle u, \Phi_{z,\xi}\rangle &= 2\langle u, \frac{\partial}{\partial \bar{z}}\Phi_{z,\xi}\rangle \\
&= -\langle u, \left(\frac{\partial}{\partial x_{n-1}} + i\frac{\partial}{\partial x_n}\right)\Phi_{z,\xi}\rangle \\
&= \langle \left(\frac{\partial}{\partial x_{n-1}} + i\frac{\partial}{\partial x_n}\right)u, \Phi_{z,\xi}\rangle = \mathcal{O}(|\xi|^{-N}),
\end{aligned}$$

for every $N > 0$, uniformly when

(10.11) $\quad z \in W_\delta$, $|x_j| \leq a-\delta$, $\left|\dfrac{\xi_k}{\xi_1}\right| \leq a-\delta$, $\xi_1 \geq 1$, $3 \leq j \leq n$, $2 \leq k \leq n$,

in view of the hypothesis on $WF((D_{x_{n-1}} + iD_{x_n})u)$. Here we have put $W_\delta = \{z \in X \,;\, \text{dist}(z, \complement W) > \delta\}$.

We now assume that $(0, \eta_0) \notin WF(u)$. If we choose a, δ above sufficiently small, then for some $\varepsilon > 0$ we have

(10.12) $\qquad\qquad\qquad \langle u, \Phi_{z,\xi}\rangle = \mathcal{O}(|\xi|^{-N})$

uniformly for each N, when $|z| < \varepsilon$ and (z, x, ξ) belongs to the domain (10.11). Inverting $\dfrac{\partial}{\partial \bar{z}}$ by means of the standard elementary solution, we can find $w(z, x_1, \ldots, x_{n-2}, \xi)$ defined in the domain (10.11) and of uniform rapid decrease in $|\xi|$, such that

$$z \mapsto \langle u, \Phi_{z,\xi}\rangle - w(z, x_1, \ldots, x_{n-2}, \xi)$$

is holomorphic with respect to z. This function is also rapidly decreasing for $|z| < \varepsilon$ and since it is also of temperate growth $\mathcal{O}(|\xi|^{N_0})$ in the domain (10.11), we can apply the Hadamard 3-circle theorem repeatedly to conclude that it is of uniform rapid decrease on the domain (10.11) if z belongs to any fixed compact subset of W_δ, provided of course that W_δ is connected. (Naturally we choose the geometry of W so that the last property is fulfilled.) Here we can choose $\delta > 0$ sufficiently small and we conclude that

$$\{(0, \ldots, 0, x_{n-1}, x_n\,;\, 1, 0, \ldots, 0)\,;\, (x_{n-1}, x_n) \in W\} \cap WF(u) = \emptyset.$$

This completes the proof of Theorem 10.4. $\qquad\square$

Exercises

Exercise 10.1

Let X be an open subset of $\mathbb{R}^n$, $n \geq 2$ and $P \in L^1_{\text{cl}}(X)$. Let the principal symbol $p(x, \xi)$ satisfy

$$p(x_0, \xi_0) = 0, \quad \frac{1}{2i}\{p, \bar{p}\}(x_0, \xi_0) < 0,$$

where (x_0, ξ_0) is some point in $T^*X\backslash 0$.

a) Show that $p = 0$ defines near (x_0, ξ_0) a smooth conic submanifold Σ of $T^*X\backslash 0$ which is symplectic (i.e. $\sigma|_\Sigma$ is non-degenerate).

b) Show that there exists $a \in C^\infty(T^*X\backslash 0)$ positively homogeneous of degree $-\frac{1}{2}$ and non-vanishing, such that $\frac{1}{2i}\{ap, \bar{a}\bar{p}\} = -1$ near (x_0, ξ_0).

c) Show that there exists $0 < b \in C^\infty(T^*X\backslash 0)$, positively homogeneous of degree 1, such that $\{\mathrm{Re}(ap), b\} = \{\mathrm{Im}(ap), b\} = 0$ near (x_0, ξ_0), and $b(x_0, \xi_0) = 1$.

d) Show that there is a homogeneous canonical transformation $\mathcal{H}$ from a conical neighborhood of $(0\,;\,(0,\ldots,0,1,0)) \in T^*\mathbb{R}^n\backslash 0$ onto a conical neighborhood of (x_0, ξ_0), such that $(b^{\frac{1}{2}} ap) \circ \mathcal{H} = \xi_n + i\,\xi_{n-1}\,x_n$.
Hint : Write
$$b^{\frac{1}{2}} ap = (b^{\frac{1}{2}} \mathrm{Re}\, ap) + ib(b^{-\frac{1}{2}} \mathrm{Im}\, ap)$$
$$= \Xi_n + i\, \Xi_{n-1} X_n$$
and compute the Poisson brackets between the functions Ξ_n, Ξ_{n-1}, X_n.

e) Discuss to what extent P can be reduced to $D_{x_n} - i\,x_n\,D_{x_{n-1}}$.

Notes

Theorem 10.1 was proved by Egorov [E] and is one of the basic results in microlocal analysis. For the reductions to the operators D_{x_n} and $D_{x_n} + i\,D_{x_{n-1}}$ we have followed Duistermaat–Hörmander [DHö], where also Theorem 10.4 was proved. The canonical form in Exercise 10.1 was obtained by Duistermaat–Sjöstrand [DS] following an earlier result in the analytic category by Sato–Kawai–Kashiwara [SaKK]. Reduction to canonical models by means of Fourier integral operators is an important method and all the developments cannot be described here. Here is an incomplete list of references where such models appear : [Ha], [I2,3], [Co1], [Me], [HS].

11 Global theory of Fourier integral operators

The aim of this chapter is to give the basic ingredients of this theory, and for simplicity we shall work on open sets in Euclidean space rather than on manifolds. (The extension to manifolds will appear to be fairly obvious.)

Let $X \subset \mathbb{R}^n$ be open and let $V \subset X \times \dot{\mathbb{R}}^N$ be conic and open.

Definition 11.1 A real-valued function $\varphi = \varphi(x, \theta) \in C^\infty(V)$ is a non-degenerate phase function if

1) φ is positively homogeneous of degree 1 in θ.

2) $d_{(x,\theta)}\varphi \neq 0$ everywhere.

3) If $\varphi'_\theta(x, \theta) = 0$, then $d\dfrac{\partial \varphi}{\partial \theta_1}, \ldots, d\dfrac{\partial \varphi}{\partial \theta_N}$ are linearly independent at (x, θ).

Let φ be a non-degenerate phase function. Then $C_\varphi = \{(x, \theta) \in V ; \varphi'_\theta(x, \theta) = 0\}$ is a closed conic submanifold of V, of codimension N, i.e. of dimension n. We observe that 3) can be reformulated by saying that $(\varphi''_{\theta x} \; \varphi''_{\theta\theta})$ should be surjective everywhere on C_φ (or equivalently that $\begin{pmatrix} \varphi''_{x\theta} \\ \varphi''_{\theta\theta} \end{pmatrix}$ should be injective on C_φ). Here $\varphi''_{\theta x}$ denotes the matrix $(\varphi''_{\theta_j x_k})_{\substack{1 \leq j \leq N \\ 1 \leq k \leq n}}$ and similarly for $\varphi''_{x\theta}, \varphi''_{\theta\theta}$.

We now consider the maps :

$$j : C_\varphi \ni (x, \theta) \mapsto (x, \varphi'_x(x, \theta)) \in T^*X \setminus 0.$$

Lemma 11.2 dj is injective at every point of C_φ.

Proof: Let $(\delta_x, \delta_\theta) \in TC_\varphi$ at some point (x, θ) so that $\varphi''_{\theta x} \delta_x + \varphi''_{\theta\theta} \delta_\theta = 0$. We have $dj(\delta_x, \delta_\theta) = (\delta_x, \varphi''_{xx} \delta_x + \varphi''_{x\theta} \delta_\theta)$, so if $dj(\delta_x, \delta_\theta) = 0$, we get first $\delta_x = 0$ and then

$$\begin{pmatrix} \varphi''_{x\theta} \\ \varphi''_{\theta\theta} \end{pmatrix} \delta_\theta = 0.$$

Hence $\delta_\theta = 0$ by 3). □

After shrinking V around any fixed point $(x_0, \theta_0) \in C_\varphi$, we may assume that j is injective and that $j(C_\varphi) \stackrel{\text{def}}{=} \Lambda_\varphi$ is a smooth conic submanifold of $T^*X \setminus 0$ of dimension n. Moreover $j : C_\varphi \to \Lambda_\varphi$ becomes a diffeomorphism.

Lemma 11.3 Λ_φ is a conic Lagrangian manifold.

Proof: We identity C_φ and Λ_φ by means of j. Then with ω denoting the canonical 1-form

$$\omega|_{\Lambda_\varphi} \simeq \sum_1^n \frac{\partial \varphi}{\partial x_j} dx_j|_{C_\varphi} = \Big(\sum \frac{\partial \varphi}{\partial x_j} dx_j + \sum \frac{\partial \varphi}{\partial \theta_k} d\theta_k\Big)\Big|_{C_\varphi} =$$

$$= d\varphi|_{C_\varphi} = d(\varphi|_{C_\varphi}) = 0,$$

since $\varphi|_{C_\varphi} = 0$ by the Euler homogeneity relation : $\varphi = \sum \frac{\partial \varphi}{\partial \theta_k} \theta_k$. □

Remark : The arguments above are valid also in the case when X is a smooth manifold.

Remark : If we drop the homogeneity assumption on φ, we can still define the smooth manifold Λ_φ and Λ_φ is still Lagrangian. In fact, $\omega|_{\Lambda_\varphi} \simeq d(\varphi|_{C_\varphi})$ and hence $\sigma = d\omega|_{\Lambda_\varphi} \simeq d^2(\varphi|_{C_\varphi}) = 0$.

Proposition 11.4 *Let $\Lambda \subset T^*X\backslash 0$ be a closed conic Lagrangian manifold. If $(x_0, \xi_0) \in \Lambda$, then we can find a non-degenerate phase function φ, such that $\Lambda = \Lambda_\varphi$ in a neighborhood of (x_0, ξ_0).*

Proof: After a change of coordinates in X we may assume that Λ is given by $-x = \frac{\partial H(\xi)}{\partial \xi}$ in a conic neighborhood of (x_0, ξ_0), where H is smooth, real-valued and positively homogeneous of degree 1 (defined in a conic neighborhood of ξ_0). Put $\varphi(x, \theta) = x\theta + H(\theta)$. Then $\varphi(x, \theta)$ is a non-degenerate phase function, $C_\varphi : x = -H'_\theta(\theta)$, $\Lambda_\varphi = \{(x, \theta) ; x = -H'_\theta(\theta)\}$. □

Let $\varphi \in C^\infty(V)$ be a non-degenerate phase function, $(x_0, \theta_0) \in C_\varphi$. Choosing the local coordinates $x_1, \ldots, x_n$ conveniently, and shrinking V around (x_0, θ_0), we may assume that Λ_φ is of the form $-x = \frac{\partial H(\xi)}{\partial \xi}$, $\xi \in W$, with W a conic neighborhood of ξ_0, and with H as in the proof of the proposition. Then it is easy to check that $\det \varphi'' \neq 0$ (possibly after shrinking V further).

Let $a \in S^{m+(n-2N)/4}(X \times \mathbb{R}^N)$ (where no subscript means that the symbols are of type 1,0) have its support in a cone $\subset\subset V$. Consider

$$I(a, \varphi) = \int e^{i\varphi(x,\theta)} a(x, \theta) d\theta.$$

The Fourier transform is then given by

$$\widehat{I(a, \varphi)}(\xi) = \int e^{i(\varphi(x,\theta) - x\xi)} a(x, \theta) d\theta.$$

For $\xi \in W$ the phase $\varphi - x\xi$ has a non-degenerate critical point $(x(\xi), \theta(\xi))$, and we see that this point is the inverse image in C_φ under j_φ, of the point $\left(-\frac{\partial H}{\partial \xi}(\xi), \xi\right) \in \Lambda_\varphi$. The corresponding critical value is $-x(\xi) \cdot \xi = H(\xi)$.

Using integrations by parts in (x, θ), we show :

1) $\widehat{I(a,\varphi)}(\xi)$ is rapidly decreasing, when $|\xi| \to \infty$, $\xi \notin W$.

2) For $\xi \in W$ we get, up to a rapidly decreasing term :

$$\widehat{I(a,\varphi)}(\xi) \equiv \iint e^{i(\varphi(x,\theta) - x\xi)} a(x,\theta) \chi\left(\frac{\theta}{|\xi|}\right) dx\, d\xi,$$

where $\chi \in C_0^\infty(\mathring{\mathbb{R}}^N)$ is equal to 1 for $\frac{1}{C} \leq |\theta| \leq C$ and $C > 0$ is sufficiently large.

Replacing $\xi \in W$ by $\lambda \xi$ with $\frac{1}{2} \leq |\xi| \leq 2$, we get after a change of variables in θ

$$\widehat{I(a,\varphi)}(\lambda\xi) = \lambda^N \iint e^{i\lambda(\varphi(x,\theta) - x\xi)} a(x, \lambda\theta) \chi\left(\frac{\theta}{|\xi|}\right) dx\, d\theta$$

and by the method of stationary phase

$$\widehat{I(a,\varphi)}(\lambda\xi) = \lambda^{\frac{N}{2} - \frac{n}{2}} e^{-i\lambda H(\xi)} A(\xi, \lambda),$$

where A is of class $S^{m - \frac{n}{4}}$ (with ξ as the base variables and λ as the fiber variables). Modulo a symbol of order $m - \frac{n}{4} - 1$, we have :

$$A \equiv A_0(\xi, \lambda) = (2\pi)^{-\frac{n+N}{2}} |\det \varphi''(x(\xi), \theta(\xi))|^{-\frac{1}{2}} e^{i\frac{\pi}{4} \operatorname{sgn} \varphi''} a(x(\xi), \lambda\theta(\xi)).$$

We notice that $\det \varphi''(x, \theta)$ is homogeneous of degree $n - N$ in θ, and we can reformulate the result just obtained (with a new symbol A) as

$$\widehat{I(a,\varphi)}(\xi) = e^{-iH(\xi)} A(\xi),$$

where $A \in S^{m - \frac{n}{4}}(\mathbb{R}^n)$ is of class $S^{-\infty}$ outside W, and inside W given modulo $S^{m - \frac{n}{4} - 1}$ by

$$(2\pi)^{-\frac{n+N}{2}} |\det \varphi''(x(\xi), \theta(\xi))|^{-\frac{1}{2}} e^{i\frac{\pi}{4} \operatorname{sgn} \varphi''} a(x(\xi), \theta(\xi)).$$

Theorem 11.5 Let $\tilde{V} \subset X \times \mathring{\mathbb{R}}^{\tilde{N}}$ and $\tilde{\varphi} \in C^\infty(\tilde{V})$ have the same properties as (V, φ) above. We also assume that $(x_0, \tilde{\theta}_0) \in C_{\tilde{\varphi}}$ is mapped by (the diffeomorphism) $j_{\tilde{\varphi}}$ to (x_0, ξ_0) and that $\Lambda_{\tilde{\varphi}} = \Lambda_\varphi$. Then for every conic neighborhood $\tilde{U} \subset \tilde{V}$ of $(x_0, \tilde{\theta}_0)$, there exists a conic neighborhood $U \subset V$, such that if $a \in S^{m+(n-2N)/4}(X \times \mathbb{R}^N)$ has its support in U, then there exists $\tilde{a} \in S^{m+(n-2\tilde{N})/4}(X \times \mathbb{R}^{\tilde{N}})$ with support in $\tilde{U}$, such that $I(a, \varphi) \equiv I(\tilde{a}, \tilde{\varphi})$ mod C^∞.

Proof: We have already seen that $\widehat{I(a, \varphi)}(\xi)$ and $\widehat{I(\tilde{a}, \tilde{\varphi})}(\xi)$ have the same general form $e^{-iH(\xi)} A(\xi)$ and $e^{-iH(\xi)} \tilde{A}(\xi)$ respectively, so in order to prove the theorem it is enough to show that we can obtain arbitrary symbols modulo $S^{-\infty}$ with support in a small conic neighborhood of ξ_0 :

Lemma 11.6 *Let $U \subset\subset V$ be a conic neighborhood of (x_0, θ_0). Then there exists a conic neighborhood W of ξ_0, such that if $A \in S^{m-\frac{n}{4}}(\mathbb{R}^n)$ has its support in W, then there exists $a \in S^{m+\frac{n}{4}-\frac{N}{2}}(X \times \mathbb{R}^N)$ with support in U, such that $\widehat{I(a, \varphi)}(\xi) - e^{-iH(\xi)} A(\xi)$ is rapidly decreasing, when $\xi \to \infty$.*

Proof of the lemma : We construct a as an asymptotic sum : $a \sim a_0 + a_1 + \ldots$, $a_j \in S^{m+\frac{n}{4}-\frac{N}{2}-j}$. First we can find a_0 such that

$$(2\pi)^{-\frac{n+N}{2}} |\det \varphi''|^{-\frac{1}{2}} e^{i\frac{\pi}{4} \operatorname{sgn} \varphi''} a_0(x(\xi), \theta(\xi)) = A(\xi).$$

Then $e^{-iH} A - \widehat{I(a_0, \varphi)}(\xi) = e^{-iH(\xi)} A_1$, where $A_1 \in S^{m+\frac{n}{4}-1}$, and we can introduce further corrections iteratively. □

The proof of the theorem is now clear. □

We can now give the global definition of Fourier integral distributions.

Definition : Let $\Lambda \subset T^*X \setminus 0$ be a closed conic Lagrangian manifold and let $m \in \mathbb{R}$. We define $I^m(X, \Lambda)$ to be the space of all $u \in \mathcal{D}'(X)$ such that

1) $WF(u) \subset \Lambda$.

2) If $(x_0, \xi_0) \in \Lambda$ and if $\varphi \in C^\infty(V)$, with V an open cone in $X \times \dot{\mathbb{R}}^N$, is a non-degenerate phase function such that $\Lambda_\varphi = \Lambda$ in a neighborhood of (x_0, ξ_0), then there exists $a \in S^{m+\frac{n}{4}-\frac{N}{2}}(X \times \mathbb{R}^N)$ with support in a cone $\subset\subset V$, such that $u \equiv I(a, \varphi)$ microlocally near (x_0, ξ_0).

Remark : This definition is easy to extend to the case when X is a paracompact manifold.

Remark : If $P \in L^{m_1}(X)$ (of type 1,0) is properly supported and if $u \in I^{m_2}(X, \Lambda)$, then $Pu \in I^{m_1+m_2}(X, \Lambda)$, as can be seen by using the asymptotic expansions at the end of Chapter 8. Using a pseudodifferential partition of unity, we can then prove that if $u \in \mathcal{D}'(X)$, then u belongs to $I^m(X, \Lambda)$ if and only if u is equal to a smooth function plus a locally finite sum of elements $I(a,\varphi)$ as above, with $\Lambda_\varphi \subset \Lambda$.

We shall next briefly discuss the notion of *principal symbol* of Fourier integral distributions. Let $\varphi(x,\theta)$ be a non-degenerate phase function. We then define an n-form d_φ on C_φ by :

$$d_\varphi \wedge d\left(\frac{d\varphi}{\partial\theta_1}\right) \wedge \ldots \wedge d\left(\frac{\partial\varphi}{\partial\theta_N}\right) = dx_1 \wedge \ldots \wedge dx_n \wedge d\theta_1 \wedge \ldots \wedge d\theta_N.$$

More explicitly, let $\lambda_1,\ldots,\lambda_n$ be local coordinates on C_φ, and extend them to smooth functions defined in a full neighborhood in $X \times \mathbb{R}^N$ of some point in C_φ. Then we get $d_\varphi = f\, d\lambda_1 \wedge \ldots \wedge d\lambda_n$, with

$$f = \frac{dx_1 \wedge \ldots \wedge dx_n \wedge d\theta_1 \wedge \ldots \wedge d\theta_N}{d\lambda_1 \wedge \ldots \wedge d\lambda_n \wedge d\frac{\partial\varphi}{\partial\theta_1} \wedge \ldots \wedge d\frac{\partial\varphi}{\partial\theta_N}} =$$

$$= \left(\det \begin{pmatrix} \dfrac{\partial\lambda}{\partial x} & \dfrac{\partial\lambda}{\partial\theta} \\ \dfrac{\partial^2\varphi}{\partial x\,\partial\theta} & \dfrac{\partial^2\varphi}{\partial\theta^2} \end{pmatrix}\right)^{-1}.$$

The definition does not depend on the choice of $\lambda_1,\ldots,\lambda_n$, but it does depend on the choice of local coordinates $x_1,\ldots,x_n$. Let us neglect the latter fact for the moment, and choose local coordinates such that Λ_φ becomes $-x = \dfrac{\partial H}{\partial \xi}$ (near some given point). We can then identity C_φ with Λ_φ and choose $\xi_1,\ldots,\xi_n$ as local coordinates. Since $\xi_j = \dfrac{\partial\varphi}{\partial x_j}$, it is natural to choose $\lambda_j = \dfrac{\partial\varphi}{\partial x_j}$. We then get

$$d_\varphi = (\det \varphi'')^{-1}\, d\xi_1 \wedge \ldots \wedge d\xi_n.$$

Instead of considering forms of maximal degree we now consider the associated densities and change the definition of d_φ into

$$d_\varphi = |\det \varphi''|^{-1}\, |d\xi_1 \wedge \ldots \wedge d\xi_n|.$$

(We recall that a density on a manifold obeys almost the same transformation rules as forms of maximal degree, under changes of local coordinates. The only difference is that we take the absolute value of all the Jacobians.)

Let $m_\lambda : \Lambda_\varphi \ni (x,\xi) \mapsto (x,\lambda\xi) \in \Lambda_\varphi$, $\lambda > 0$. If ω is an n-form or a density on Λ_φ, we say that ω is homogeneous of degree m, if $m_\lambda^*\omega = \lambda^m\omega$. The form $d\xi_1 \wedge \ldots \wedge d\xi_n$ is homogeneous of degree n, and since $|\det \varphi''|$ is homogeneous of degree $n - N$, we conclude that the density d_φ is homogeneous of degree N.

In order to keep better track of the powers of 2π, we write from now on

$$(11.1) \qquad I(a,\varphi) = (2\pi)^{-(n+2N)/4} \int e^{i\varphi(x,\theta)} a(x,\theta)\, d\theta.$$

In the following discussion we restrict our attention to classical symbols. This is for simplicity only, and symbols of type 1,0 or even $\rho, 1 - \rho$ with $\rho > \frac{1}{2}$ could easily be handled. Let $a \in S_{\text{cl}}^{m+\frac{n}{4}-\frac{N}{2}}$ and let $a^0 = a_{m+\frac{n}{4}-\frac{N}{2}}$ be the leading homogeneous part of a. Then the $\frac{1}{2}$-density $a^0 \sqrt{d_\varphi}$ is homogeneous of degree $m + \frac{n}{4}$. (For a change of local coordinates, if the transformation rule for forms of maximal degree is given by multiplication by a Jacobian J, then for $\frac{1}{2}$-densities the transformation rule is to multiply by $|J|^{\frac{1}{2}}$.)

Proposition 11.7 *Let* $\tilde{a} \in S_{\text{cl}}^{m+\frac{n}{4}-\frac{\tilde{N}}{2}}(X \times \mathbb{R}^{\tilde{N}})$ *and* $\tilde{\varphi}(x,\tilde{\theta})$ *a corresponding non-degenerate phase function and assume that* $\Lambda_\varphi = \Lambda_{\tilde{\varphi}}$ *near a given point* (x_0,ξ_0). *Finally assume that* $I(a,\varphi) \equiv I(\tilde{a},\tilde{\varphi})$ *microlocally near* (x_0,ξ_0). *Then near that point, we have*

$$(11.2) \qquad \tilde{a}^0 \sqrt{d_{\tilde{\varphi}}} = a^0 \sqrt{d_\varphi} \cdot e^{i\frac{\pi}{4}(\operatorname{sgn}\varphi'' - \operatorname{sgn}\tilde{\varphi}'')}.$$

We leave the proof of this proposition as well as of the next one as an exercise.

Proposition 11.8 *If we consider* $I(a,\varphi)$ *as a* $\frac{1}{2}$-*density distribution, then* $a^0 \sqrt{d_\varphi}$ *is independent of the choice of local coordinates.*

Proposition 11.9 *If* $a^0 \sqrt{d_\varphi} = 0$ *(or in other words* $a^0|_{C_\varphi} = 0$*), then there exists* $\tilde{a} \in S_{\text{cl}}^{m+\frac{n}{4}-\frac{N}{2}-1}$, *such that* $I(a,\varphi) \equiv I(\tilde{a},\varphi) \mod C^\infty$.

Proof: This result is essentially obvious from the earlier characterization of Fourier integral distributions by means of Fourier transforms, but a direct proof may be instructive: using Taylor's formula and a partition of unity, we may write

$$a = \sum_{1}^{N} b_j(x,\theta) \frac{\partial \varphi}{\partial \theta_j} + c$$

with $b_j \in S_{\text{cl}}^{m+\frac{n}{4}-\frac{N}{2}}$, $c \in S_{\text{cl}}^{m+\frac{n}{4}-\frac{N}{2}-1}$.

It then suffices to notice that

$$\int e^{i\varphi} \frac{\partial \varphi}{\partial \theta_j} b_j \, d\theta = \frac{1}{i} \int \frac{\partial}{\partial \theta_j} (e^{i\varphi}) b_j \, d\theta = -\frac{1}{i} \int e^{i\varphi} \frac{\partial b_j}{\partial \theta_j} \, d\theta$$

and that $\frac{\partial b_j}{\partial \theta_j} \in S_{\text{cl}}^{m+\frac{n}{4}-\frac{N}{2}-1}$. □

Return to (11.2). We define the Maslov (or Maslov–Keller–Hörmander) line bundle $\mathcal{L}$ over Λ as the complex line bundle equipped with the transition functions $e^{i\frac{\pi}{4}(\operatorname{sgn}\varphi'' - \operatorname{sgn}\psi'')}$ in $\Lambda_\varphi \cap \Lambda_\psi$ associated to a change of phase function and $e^{i\frac{\pi}{4}(\operatorname{sgn}\varphi''_{(x,\theta),(x,\theta)} - \operatorname{sgn}\varphi''_{(y,\theta),(y,\theta)})}$ in Λ_φ associated to a change of local coordinates. To $I(a,\varphi)$ in (11.1), we associate a section in $\Omega_{\frac{1}{2}} \otimes \mathcal{L}$, where $\Omega_{\frac{1}{2}}$ denotes the line bundle of $\frac{1}{2}$-densities on Λ, which for given local coordinates x, is given by $a^0 \sqrt{d_\varphi}$. Notice that the transition functions of $\mathcal{L}$ are of the form $e^{i\frac{\pi}{4}k}$, $k \in \mathbb{Z}$. Introducing a factor $e^{i\frac{\pi}{4}N}$, we may even reduce the transition functions to the form $e^{i\frac{\pi}{2}k}$, since when the numbers of θ (resp. $\tilde{\theta}$) variables differ by an even number, we have $\operatorname{sgn}\varphi'' - \operatorname{sgn}\tilde{\varphi}'' \in 2\mathbb{Z}$, $\operatorname{sgn}\varphi''_{(y,\theta),(y,\theta)} - \operatorname{sgn}\tilde{\varphi}''_{(x,\theta),(x,\theta)} \in 2\mathbb{Z}$.

We have essentially proved

Theorem 11.10 *We have a bijective map*

$$I_{\text{cl}}^m(X,\Lambda)/I_{\text{cl}}^{m-1}(X,\Lambda) \longrightarrow \Gamma^{m+\frac{n}{4}}(\Lambda, \Omega_{\frac{1}{2}} \otimes \mathcal{L}).$$

Here $\Gamma^j(\Lambda, \Omega_{\frac{1}{2}} \otimes \mathcal{L})$ denotes the space of smooth sections $\Omega_{\frac{1}{2}} \otimes \mathcal{L}$ which are homogeneous of degree j.

Let $P \in L_{\text{cl}}^m(X)$ be properly supported and view P as acting on $\frac{1}{2}$-densities on X. Then the transformation rules for the full symbol of P under changes of local coordinates change, but we see that the principal symbol p of P is still a well defined homogeneous function on $T^*X \backslash 0$. If $u \in I_{\text{cl}}^k(X,\Lambda)$ has the principal symbol $\alpha \in \Gamma^{k+\frac{n}{4}}(\Lambda, \Omega_{\frac{1}{2}} \otimes \mathcal{L})$, then $Pu \in I_{\text{cl}}^{m+k}(X,\Lambda)$ has the principal symbol $(p|_\Lambda)\alpha$. In WKB constructions we frequently have (by the very construction of Λ) $p|_\Lambda = 0$. Then $Pu \in I_{\text{cl}}^{m+k-1}(X,\Lambda)$, and if we want to construct u with Pu of still lower order, we expect to have to solve a transport equation for α. The next theorem gives an invariant version of the transport equation, involving the important *subprincipal symbol* of P.

Theorem 11.11 *Let $P \in L_{\text{cl}}^m(X)$ be a properly supported pseudodifferential operator acting on $\frac{1}{2}$-densities. This means that for every choice of local coordinates $x_1, \ldots, x_n$ we can write the $\frac{1}{2}$-densities as $f(x)(dx_1 \ldots dx_n)^{\frac{1}{2}}$*

and view P as a pseudodifferential operator in the usual sense, acting on the functions f. We then have a corresponding complete symbol $\sim p_m(x,\xi) + p_{m-1}(x,\xi) + \ldots$, with p_j positively homogeneous of degree j.

(A) The subprincipal symbol $S_P(x,\xi) = p_{m-1}(x,\xi) - \dfrac{1}{2i} \sum \dfrac{\partial^2 p_m}{\partial x_j \, \partial \xi_j}$, which is positively homogeneous of degree $m-1$ on $T^*X\backslash 0$, does not depend on the choice of local coordinates.

(B) Let $u \in I_{\text{cl}}^k(X,\Lambda)$ have the principal symbol α and assume that $p = p_m$ vanishes on Λ. Then the principal symbol of $P(u) \in I_{\text{cl}}^{m+k-1}(X,\Lambda)$ is equal to $\left(\frac{1}{i}\mathcal{L}_{H_p} + S_P\right)\alpha$, where $\mathcal{L}_{H_p}$ denotes the Lie derivative with respect to the Hamilton field H_p.

We refrain from giving a proof, which in principle consists of a more or less straightforward computation.

We end this chapter with a brief account of the global theory of Fourier integral operators and in particular we explain how to compose such operators. Let $X \subset \mathbb{R}^{n_X}$, $Y \subset \mathbb{R}^{n_Y}$ be open (or more generally paracompact smooth) manifolds. If $K \in \mathcal{D}'(X \times Y; \Omega_{X\times Y}^{\frac{1}{2}})$ is a $\frac{1}{2}$-density valued distribution, we can (by the Schwartz kernel theorem) define a corresponding operator $A : C_0^\infty(Y; \Omega_Y^{\frac{1}{2}}) \to \mathcal{D}'(X; \Omega_X^{\frac{1}{2}})$ and conversely to every such A there is a unique associated kernel K as above.

Definition: A Lagrangian manifold $C \subset T^*(X \times Y) \simeq T^*X \times T^*Y$ for the symplectic form $\sigma_X - \sigma_Y$ (cf. the discussion of canonical transformations in Chapter 9) is called a canonical relation.

Equivalently we can say that $C \subset T^*X \times T^*Y$ is a canonical relation if

$$C' \stackrel{\text{def}}{=} \{(x,\xi\,;\,y,-\eta)\,;\,(x,\xi\,;\,y,\eta) \in C\}$$

is a Lagrangian manifold for the standard symplectic form $\sigma_{X\times Y} = \sigma_X + \sigma_Y$.

Let $C \subset T^*(X \times Y)\backslash 0$ be a closed conic canonical relation, contained in $(T^*X\backslash 0) \times (T^*Y\backslash 0)$. A Fourier integral operator of order m associated to C is by definition an operator A acting on $\frac{1}{2}$-densities as above, where the distribution kernel K belongs to $I^m(X \times Y, C')$. By abuse of notation, we shall write $A \in I^m(X \times Y, C')$. Both the closedness assumption on C and the assumption that $C \subset (T^*X\backslash 0) \times (T^*Y\backslash 0)$ may be relaxed, if we instead make a restriction on $WF'(A)$. With this in mind, we check that the definition just given generalizes those of Chapter 10, where C was a graph of a canonical transformation. If $X = Y$ and C is the graph of $\mathcal{H} = \text{id}$, then the corresponding space of Fourier integral operators coincides with the space of classical pseudodifferential operators.

Since $C \subset (T^*X\backslash 0) \times (T^*Y\backslash 0)$ we know from Chapter 7 that every corresponding Fourier integral operator maps $C_0^\infty(Y; \Omega_{\frac{1}{2}}) \to C^\infty(X; \Omega_{\frac{1}{2}})$, $\mathcal{E}'(Y; \Omega_{\frac{1}{2}}) \to \mathcal{D}'(X; \Omega_{\frac{1}{2}})$.

Let X, Y, Z be smooth paracompact manifolds and let $C_1 \subset T^*(X \times Y)\backslash 0$, $C_2 \subset T^*(Y \times Z)\backslash 0$ be closed conic canonical relations contained in $(T^*X\backslash 0) \times (T^*Y\backslash 0)$ and $(T^*Y\backslash 0) \times (T^*Z\backslash 0)$ respectively.

Assume:

(a) $C_1 \times C_2$ and $\tilde{\Delta} \stackrel{\text{def}}{=} (T^*X\backslash 0) \times \text{diag}((T^*Y\backslash 0) \times (T^*Y\backslash 0)) \times (T^*Z\backslash 0)$ intersect transversally (in the sense that the sum of the tangent spaces is equal to the full tangent space of $(T^*X\backslash 0) \times (T^*Y\backslash 0)^2 \times (T^*Z\backslash 0)$) at every point of intersection.

(b) The natural projection $C_1 \times C_2 \cap \tilde{\Delta} \to T^*(X \times Z)\backslash 0$ is injective and proper.

Theorem 11.12 *Under the above assumptions, $C_1 \circ C_2 \subset T^*(X \times Z)\backslash 0$ (i.e. the image of the map in (b)) is a closed conic canonical transformation, contained in $(T^*X\backslash 0) \times (T^*Z\backslash 0)$. If $A_1 \in I_{cl}^{m_1}(X \times Y; C_1')$, $A_2 \in I_{cl}^{m_2}(Y \times Z; C_2')$ are Fourier integral operators with at least one of them properly supported, then $A_1 \circ A_2 \in I_{cl}^{m_1+m_2}(X \times Z; (C_1 \circ C_2)')$.*

Idea of the proof: We will only discuss the microlocal situation. Let $(x_0, \xi_0; y_0, \eta_0) \in C_1$, $(y_0, \eta_0; z_0, \zeta_0) \in C_2$, and let $\varphi_1(x, y, \theta)$, $\varphi_2(y, z, w)$ be non-degenerate phase functions, generating $\Lambda_1 = C_1'$ and $\Lambda_2 = C_2'$ near $(x_0, \xi_0; y_0, -\eta_0)$ and $(y_0, \eta_0; z_0, -\zeta_0)$ respectively. Let

$$\tilde{N}^* = (T^*X\backslash 0) \times N^*(\text{diag}(Y \times Y)) \times (T^*Z\backslash 0)$$

where the middle factor is the conormal bundle of $\text{diag}(Y \times Y)$ in $Y \times Y$. Then (a) implies the following sequence of equivalent properties:

- $\Lambda_1 \times \Lambda_2$ and $\tilde{N}^*$ intersect transversally at $(x_0, \xi_0; y_0, -\eta_0; y_0, \eta_0; z_0, -\zeta_0)$.

- The differential of the map
$$\Lambda_1 \times \Lambda_2 \ni (x, \xi; y, \eta; y', \eta'; z, \zeta) \mapsto (y - y', \eta + \eta')$$
is surjective at $(x_0, \xi_0; y_0, -\eta_0; y_0, \eta_0; z_0, -\zeta_0)$.

- The differential of the map
$$C_{\varphi_1} \times C_{\varphi_2} \ni (x, y, \theta; y', z, w) \mapsto \left(y - y', \frac{\partial \varphi_1}{\partial y} + \frac{\partial \varphi_2}{\partial y'}\right)$$
is surjective at $(x_0, y_0, \theta_0; y_0, z_0, w_0)$, where $(x_0, y_0, \theta_0) \in C_{\varphi_1}$, $(y_0, z_0, w_0) \in C_{\varphi_2}$ are the points corresponding to $(x_0, \xi_0; y_0, -\eta_0)$ and $(y_0, \eta_0; z_0, -\zeta_0)$ respectively.

- The differential of

$$X \times Y \times \mathbb{R}^{N_\theta} \times Y \times Z \times \mathbb{R}^{N_w} \ni (x,y,\theta; y',z,w) \mapsto$$

$$\left(y - y', \frac{\partial \varphi_1}{\partial y} + \frac{\partial \varphi_2}{\partial y'}, \frac{\partial \varphi_1}{\partial \theta}, \frac{\partial \varphi_2}{\partial w}\right)$$

is surjective at $(x_0, y_0, \theta_0; y_0, z_0, w_0)$.

- The differential of

$$X \times Y \times Z \times \mathbb{R}^{N_\theta} \times \mathbb{R}^{N_w} \ni (x,y,z,\theta,w) \mapsto \left(\frac{\partial \varphi_1}{\partial y} + \frac{\partial \varphi_2}{\partial y'}, \frac{\partial \varphi_1}{\partial \theta}, \frac{\partial \varphi_2}{\partial w}\right)$$

is surjective at $(x_0, y_0, z_0, \theta_0, w_0)$.

The last property means that if we drop the homogeneity assumption in the definition of non-degenerate phase functions, then $\Phi(x,z;y,\theta,w) = \varphi_1(x,y,\theta) + \varphi_2(y,z,w)$ is such a function near $(x_0, y_0, z_0, \theta_0, w_0)$, with (y, θ, w) as the fiber variables. The corresponding Lagrangian manifold is (locally) equal to $(C_1 \circ C_2)'$, and hence $C_1 \circ C_2$ is a canonical relation.

If we compose

$$A_1 v(x) = \iint e^{i\varphi_1(x,y,\theta)} a_1(x,y,\theta) v(y) \, dy \, d\theta$$

and

$$A_2 u(y) = \iint e^{i\varphi_2(y,z,w)} a_2(y,z,w) u(z) \, dz \, dw,$$

we get

$$A_1 A_2 u(x) = \iint e^{i\Phi(x,z;y,\theta,w)} a_1(x,y,\theta) a_2(y,z,w) u(y) \, dy \, d\theta \, dz \, dw.$$

Modifying $A_1 A_2$ by a smoothing operator, we may introduce a cut-off $\chi\left(\frac{|\theta|}{|w|}\right)$ where $\chi \in C_0^\infty(]0,\infty[)$, $\chi(\lambda) = 1$ if $\varepsilon \leq \lambda \leq \frac{1}{\varepsilon}$, for some sufficiently small $\varepsilon > 0$, and then introduce the new fiber variables $\omega = \left((\theta^2 + w^2)^{\frac{1}{2}} y, \theta, w\right)$ with respect to which Φ is homogeneous of degree 1. (This does not change Λ_Φ.)

□

Exercises

Exercise 11.1

(The unitary group of the harmonic oscillator)

Let $P = \frac{1}{2}(D_x^2 + x^2)$. For t near 0, we want to solve

$$\begin{cases} (D_t + P)u = 0 \\ u|_{t=0} = v. \end{cases}$$

We try
$$u = U_t v(x) = \frac{1}{2\pi} \int e^{i\varphi(t,x,\eta)} a(t) \hat{v}(\eta) d\eta$$

with $\varphi(t,\cdot,\cdot\cdot)$ a quadratic form in (x,η).

a) Explain why it is natural to require φ to satisfy the eikonal equation

$$\frac{\partial \varphi}{\partial t} + \frac{1}{2}\left(\left(\frac{\partial \varphi}{\partial x}\right)^2 + x^2\right) = 0, \qquad \varphi|_{t=0} = x \cdot \eta.$$

b) Let $p = \frac{1}{2}(\xi^2 + x^2)$. Calculate H_p and $\exp tH_p$.

c) Let $\Lambda_{\varphi(t,\cdot,\eta)} = \{(x, \varphi_x'(t,x,\eta)); x \in \mathbb{R}\}$.

Determine $\Lambda_{\varphi(t,\cdot,\eta)}$ with the help of Hamilton–Jacobi theory and show that φ must be of the form

$$\varphi(t,x,\eta) = C(t,\eta) + \frac{\eta}{\cos t} x - \frac{\tan t}{2} x^2.$$

d) Substitute this expression into the eikonal equation and get

$$\varphi(t,x,\eta) = -\frac{\tan t}{2} x^2 + \frac{x\eta}{\cos t} - \frac{\tan t}{2} \eta^2.$$

e) Derive a transport equation for a and show that with $a(t) = (\cos t)^{-1/2}$ we get a solution $u \in C^\infty\left(\left]-\frac{\pi}{2}, \frac{\pi}{2}\right[\times \mathbb{R}\right)$ of the initial value problem if $v \in \mathcal{S}(\mathbb{R})$.

f) Check that $\varphi(t,\cdot,\cdot\cdot)$ is a generating function of the canonical transform $\exp(t H_p)$.

g) Let $\psi(x,\eta)$ be a quadratic form with $\dfrac{\partial^2 \psi}{\partial x \, \partial \eta} \neq 0$. Let κ_ψ be the associated canonical transformation. Show that if $\ell(x,\xi)$, $m(x,\xi)$ are linear forms related by $\ell \circ \kappa_\psi = m$ and if $Av(x) = \int e^{i\psi(x,\eta)} \hat{v}(\eta) d\eta$ (so that $A: \mathcal{S}(\mathbb{R}) \to \mathcal{S}(\mathbb{R})$), then $\ell(x, D_x) A = m(x, D_x)$.

Finally show that

$$U_t : \mathcal{S}(\mathbb{R}) \to C^\infty\left(\left]-\frac{\pi}{2}, \frac{\pi}{2}\right[; \mathcal{S}(\mathbb{R})\right).$$

h) Show that U_t extends as an isometry $L^2(\mathbb{R}) \to L^2(\mathbb{R})$. Show that if u_1 and u_2 are solutions in $C^\infty\left(\left]-\frac{\pi}{2}, \frac{\pi}{2}\right[; \mathcal{S}(\mathbb{R})\right)$ of the initial problem with $u_1|_{t=0} = u_2|_{t=0} = v$ then $\|u_1(t) - u_2(t)\| = 0$.

i) Show that $U_t U_s = U_{t+s}$ if $|t| < \frac{\pi}{2}$, $|s| < \frac{\pi}{2}$, $|t+s| < \frac{\pi}{2}$. Then show how one can extend the definition of U_t to all $t \in \mathbb{R}$.

j) With the help of the stationary phase method (here in an "exact version") show that for $0 < t < \pi$

$$U_t v(x) = e^{-i\frac{\pi}{4}} (\sin t)^{-1/2} \frac{1}{\sqrt{2\pi}} \int e^{i(\frac{1}{2}\frac{\cos t}{\sin t} x^2 - \frac{xy}{\sin t} + \frac{1}{2}\frac{\cos t}{\sin t} y^2)} v(y) dy.$$

In particular, $U_{\frac{\pi}{2}} v(x) = \dfrac{e^{-i\frac{\pi}{4}}}{\sqrt{2\pi}} \int e^{-ixy} v(y)\, dy$.

k) Let $\tilde{P} = P - \frac{1}{2}$ and $\tilde{U}_t = e^{i\frac{t}{2}} U_t$ be the associated unitary group. Show that $\tilde{U}_{2\pi} = \mathrm{Id}$. Show then that the eigenvalues of $\tilde{P}$ are integers. (We may assume that the eigenfunctions are in $\mathcal{S}(\mathbb{R})$.)

l) Calculate U_t for $t \in \mathbb{R}\setminus\mathbb{Z}\pi$.

Exercise 11.2

Explain how to construct $u(t, x) \in \mathcal{D}'(\mathbb{R} \times \mathbb{R}^2)$ in Exercise 6.3.

Notes

A global theory in the framework of oscillating parameter-dependent functions was given by Maslov [Ma1] (see also Leray [Le]); one of the motivations was to have a better understanding of the behavior of WKB-solutions near caustics (cf. Keller [Ke]). The theory here, more directly applicable to classical questions in partial differential equations, is due to Hörmander and we have followed [Hö2] except for the proof of the equivalence of phase functions which is taken from [MS], and which is also used in [Hö4] in a different formalism. There have been many generalizations and extensions of the theory, see for instance [Hö4], [R].

12 Spectral theory for elliptic operators

In this chapter we shall establish a theorem of Hörmander on the asymptotic behaviour of the eigenvalues of an elliptic self-adjoint operator on a compact manifold.

Let X be a compact connected (C^∞) manifold of dimension n, equipped with a strictly positive C^∞ density dx. Let P be a differential operator on X of order $m \geq 1$ with smooth coefficients. We assume that P is elliptic and formally self-adjoint with respect to dx:

$$\int Pu(x)\overline{v(x)}\,dx = \int u(x)\overline{Pv(x)}\,dx, \qquad u,v \in C^\infty(X).$$

Let $p(x,\xi) \in C^\infty(T^*X)$ be the principal symbol of P. It is a real-valued homogeneous polynomial in ξ of degree m. Since $p(x,\xi) \neq 0$ when $\xi \neq 0$, p has constant sign on each connected component of $T^*X\backslash 0$. For $n \geq 2$, $T^*X\backslash 0$ is connected and we may then assume without loss of generality that $p(x,\xi) > 0$ for $\xi \neq 0$ (and then necessarily m is even). For $n = 1$ we introduce the extra assumption that $p(x,\xi) \geq 0$.

We let $H^s = H^s(X)$ be the Sobolev space of order s and we let $\|\cdot\|_s$ denote a norm on $H^s(X)$. For $s = 0$, we have the basic Hilbert space $H^0(X) = L^2(X)$ which we equip with the scalar product $(u \mid v) = \int u(x)\overline{v(x)}\,dx$. We then have the following facts:

(12.1) If we equip P with the domain $C^\infty(X)$, P becomes an unbounded symmetric operator which is essentially self-adjoint, i.e. P has a unique self-adjoint extension (which we shall also denote by P). The domain of the self-adjoint extension is $\mathcal{D}(P) = H^m$. More generally for every $k \in \mathbb{N}$, $\mathcal{D}(|P|^k) = H^{mk}$.

(12.2) There exists a constant $C > 0$ such that $P \geq -CI$ in the sense of self-adjoint operators (see Exercise 4.8). After replacing P by $P+(C+1)I$, we may assume that $P \geq I$.

(12.3) The spectrum of P is discrete : it consists only of (real) eigenvalues, where each eigenvalue is isolated and of finite multiplicity. Let $1 \leq \lambda_1 \leq \lambda_2 \leq \ldots$ be the eigenvalues of P repeated according to their multiplicity, and let $e_1, e_2, \ldots \in H^0$ be a corresponding orthonormal basis of eigenfunctions. The map $F : H^0 \to \ell^2$, defined by $F(u)(j) = (u \mid e_j)$, $j = 1, 2, \ldots$ is unitary and gives a spectral representation of P: $FPF^{-1}(\alpha_j) = (\lambda_j \alpha_j)$.

We do not prove these elementary facts, nor the following theorem of Seeley. For a detailed proof, see the book of Shubin [Sh], or the original paper [Se].

Theorem 12.1 $P^{\frac{1}{m}}$ (defined in the sense of the theory of self-adjoint operators) belongs to $L^1_{\text{cl}}(X)$ and has the homogeneous principal symbol $p^{\frac{1}{m}}$.

Proposition 12.2 There exists an $N_0 > 0$, such that $\sum \lambda_j^{-N_0} < \infty$.

Proof: Let $Q \in L^{-m}_{\text{cl}}(X)$ be a parametrix of P. Then $PQ = I + K$, $K \in L^{-\infty}$, and hence $Q = P^{-1} + P^{-1}K$. Here $P^{-1} : H^{km} \to H^{(k+1)m}$ for every $k \in \mathbb{N}$, and hence $P^{-1}K : \mathcal{D}' \to C^\infty$ is smoothing. We conclude that $P^{-1} \in L^{-m}_{\text{cl}}$ and for $\nu \in \mathbb{N}$, we have $P^{-\nu} = P^{-1} \circ \ldots \circ P^{-1} \in L^{-m\nu}_{\text{cl}}$. If $m\nu > n$, the distribution kernel of $P^{-\nu}$ is continuous and hence $P^{-\nu}$ is a Hilbert–Schmidt operator, so that $\sum_{j=1}^{\infty} \lambda_j^{-2\nu} < \infty$. □

We shall mainly work with the self-adjoint first-order classical pseudo-differential operator $Q = P^{\frac{1}{m}}$, which has the principal symbol $q = p^{\frac{1}{m}}$. It is easy to see that the domain of Q is given by $\mathcal{D}(Q) = H^1(X)$ and more generally that $\mathcal{D}(Q^k) = H^k(X)$ for every $k \in \mathbb{N}$. The eigenvalues of Q are $\mu_j = \lambda_j^{\frac{1}{m}}$, $j = 1, 2, \ldots$.

A trace formula: For $u \in H^0$, $t \in \mathbb{R}$, we put

$$(12.4) \qquad U(t)u = \sum_{1}^{\infty} e^{it\mu_j}(u \mid e_j) e_j,$$

where the series converges in the H^0-norm. For every t, $U(t)$ is a unitary operator and we have

$$(12.5) \qquad U(0) = I, \quad U(t+s) = U(t)U(s), \qquad t, s \in \mathbb{R}.$$

For $k, \ell \in \mathbb{N}$, let $C^k(\mathbb{R}; H^\ell)$ denote the space of k times continuously differentiable vector-valued functions $\mathbb{R} \ni t \mapsto u(t) \in H^\ell$. Then for every $k \in \mathbb{N}$, we have

$$(12.6) \qquad U(t)u \in C^k(\mathbb{R}; H^0) \cap C^{k-1}(\mathbb{R}; H^1) \cap \ldots \cap C^0(\mathbb{R}; H^k)$$
for every $u \in H^k$.

For $u \in H^1(X)$, we also have

$$D_t U(t)u - QU(t)u = 0, \quad U(0)u = u,$$

so that $v(t, x) = U(t)u(x)$ is a solution to the Cauchy problem

$$(D_t - Q)v = 0, \qquad v\big|_{t=0} = u.$$

Let $\chi \in \mathcal{S}(\mathbb{R})$, and consider the operator $\int \chi(t) U(t) \, dt$, defined by

$$\left(\int \chi(t) U(t) \, dt\right) u = \int \chi(t) U(t) u \, dt, \qquad u \in H^0.$$

Here the last integral can be defined as a vector-valued Riemann integral. Clearly $\int \chi(t) U(t) \, dt$ is a bounded operator $H^0 \to H^0$ of norm $\leq \|\chi\|_{L^1}$. Applying this operator first to finite linear combinations, and then using a density argument, we get

$$\left(\int \chi(t) U(t) \, dt\right) u = \sum_1^\infty \int e^{it\mu_j} \chi(t) \, dt \, (u \mid e_j) \, e_j$$

$$= \sum_1^\infty \hat\chi(-\mu_j) (u \mid e_j) e_j.$$

Proposition 12.2 implies that for some sufficiently large $N_0 > 0$:

$$\sum \mu_j^{-N_0} < \infty.$$

The Sobolev inequalities give

$$\|e_j\|_{C^k} \leq C_k \|e_j\|_{k+n+1} \leq C_k' \|Q^{k+n+1} e_j\|_0 \leq C_k' \mu_j^{k+n+1}.$$

Hence $\|e_j(x)\overline{e_j(y)}\|_{C^k} \leq C_k'' \mu_j^{k+2(n+1)}$. On the other hand, since $\chi \in \mathcal{S}(\mathbb{R})$, we have for every $N \geq 0$:

$$|\hat\chi(\mu)| \leq C_N \mu^{-N}, \quad \mu \geq 1.$$

We conclude that $\sum \hat\chi(-\mu_j) e_j(x) \overline{e_j(y)}$ converges in $C^\infty(X \times X)$.

Let $K_\chi(x,y) \in C^\infty(X \times X)$ be the sum. Considering first the case when u is a finite linear combination of the e_j, we see that K_χ is the distribution kernel of $\int \chi(t) U(t) \, dt$:

$$\left(\int \chi(t) U(t) \, dt \, u\right)(x) = \int K_\chi(x,y) u(y) \, dy.$$

We then have the trace formula

(12.7) $$\int K_\chi(x,x) \, dx = \sum \hat\chi(-\mu_j) = \hat\chi * \mu(0),$$

where $\mu \in \mathcal{S}'(\mathbb{R})$ is the counting measure, defined by

(12.8) $$\mu = \sum \delta(t - \mu_j).$$

Approximate solution of the Cauchy problem

We consider the Cauchy problem :

(12.9)
$$\begin{cases} (D_t - Q)u = 0 \\ u|_{t=0} = v. \end{cases}$$

Using the method of Chapter 6, we can construct an operator V with the following properties : there exists an $\varepsilon_0 > 0$ such that V is continuous

(12.10) $\qquad V : C^\infty(X) \to C^\infty(]-\varepsilon_0, \varepsilon_0[\times X)$

(12.11) $\qquad V : \mathcal{D}'(X) \to C^\infty(]-\varepsilon_0, \varepsilon_0[; \mathcal{D}'(X))$

(12.12) $\quad (D_t - Q) \circ V \overset{\text{def}}{=} R_0 \quad$ belongs to $\quad I^{-\infty}(]-\varepsilon_0, \varepsilon_0[\times X \times X)$
and $Vu|_{t=0} = u, \quad \forall u \in \mathcal{D}'(X).$

More precisely, by working first in local coordinates and then globalizing by means of a partition of unity, we obtain V of the form

$$(V(t)u)(x) = Vu(t,x) = \sum_{1}^{N} \chi_j(x) \left(V_j \psi_j(\cdot) u(\cdot)\right)(t,x),$$

where $\chi_j, \psi_j \in C_0^\infty(X)$, $\chi_j = 1$ near $\operatorname{supp} \psi_j$, $\sum \psi_j = 1$, and in suitable local coordinates, depending on j :

$$V_j(t)u(x) = \frac{1}{(2\pi)^n} \iint e^{i(\varphi_j(t,x,\eta) - y \cdot \eta)} a_j(t,x,\eta) u(y) \, dy \, d\eta,$$

where $\varphi = \varphi_j$, $a = a_j$ satisfy

$$\varphi|_{t=0} = x \cdot \eta, \quad \partial_t \varphi - q(x, \varphi'_x) = 0, \quad a \in S^0_{\text{cl}}, \quad a|_{t=0} = 1.$$

Proposition 12.3 $\quad U - V \overset{\text{def}}{=} R$ belongs to $I^{-\infty}(]-\varepsilon_0, \varepsilon_0[\times X \times X)$.

Proof: Consider $W(t) = U(-t)V(t)$, where $V(t)u(x) = (Vu)(t,x)$. For $u \in H^1$, we have :

$$\begin{aligned} D_t W(t) u &= -U(-t) QV(t) u + U(-t) QV(t) u + U(-t) R_0(t) u \\ &= U(-t) R_0(t) u. \end{aligned}$$

Using (12.6) and (12.12), we see that $U(-t) R_0(t)$ is smoothing : $\mathcal{D}'(X) \to C^\infty(]-\varepsilon_0, \varepsilon_0[\times X)$, and hence belongs to $I^{-\infty}(]-\varepsilon_0, \varepsilon_0[\times X \times X)$. Since $W(0) = I$, we obtain by integrating from 0 to t, that $W(t) \equiv I \mod I^{-\infty}$

$(]-\varepsilon_0, \varepsilon_0[\times X \times X)$. Hence (using (12.6) once more) $V(t) \equiv U(t)$ mod $I^{-\infty}$ $(]-\varepsilon_0, \varepsilon_0[\times X \times X)$. □

Asymptotics of $\sum_{j=1}^{\infty} \hat{\chi}(\lambda - \mu_j)$

Let $R(t, x, y)$ be the kernel of R. Let $\chi \in C_0^\infty(]-\varepsilon_0, \varepsilon_0[)$. For $u \in C^\infty(X)$ we have

$$(12.13) \int \chi(t) U(t) u(x) \, dt$$
$$= \sum_1^N \chi_j(x) \iiint \chi(t) e^{i(\varphi_j(t,x,\eta) - y \cdot \eta)} a_j(t, x, \eta) \psi_j(y) u(y) \, dy \, \frac{d\eta}{(2\pi)^n} \, dt$$
$$+ \iint \chi(t) R(t, x, y) u(y) \, dy \, dt$$

(where the local coordinates used for each term of the sum in the RHS depend on j). The distribution kernel of $\int \chi(t) U(t) \, dt$ is therefore:

$$(12.14) \quad K_\chi(x, y) = \sum_1^N \chi_j(x) \iint \chi(t) e^{i(\varphi_j(t,x,\eta) - y \cdot \eta)} a_j(t, x, \eta) \, dt \, \frac{d\eta}{(2\pi)^n} \psi_j(y)$$
$$+ \int \chi(t) R(t, x, y) \, dt.$$

In particular,

$$(12.15) \quad \int K_\chi(x, x) \, dx =$$
$$\sum_1^N \iiint \chi(t) e^{i(\varphi_j(t,x,\eta) - x \cdot \eta)} a_j(t, x, \eta) \psi_j(x) \, dt \, dx \, \frac{d\eta}{(2\pi)^n}$$
$$+ \iint \chi(t) R(t, x, x) \, dt \, dx.$$

We now replace $\chi(t)$ by $\chi(t) e^{-it\lambda}$ for a fixed $\chi \in C_0^\infty(]-\varepsilon_0, \varepsilon_0[)$ and with $\lambda \in \mathbb{R}$, $|\lambda| \to \infty$.

The trace formula gives $\int K_{\chi e^{-i(\cdot)\lambda}}(x, x) \, dx = \sum \hat{\chi}(\lambda - \mu_j)$ so (12.15) implies

$$(12.16) \quad \sum_{j=i}^{\infty} \hat{\chi}(\lambda - \mu_j) =$$
$$\sum_i^N \iiint \chi(t) e^{i(-t\lambda + \varphi_j(t,x,\eta) - x \cdot \eta)} a_j(t, x, \eta) \psi_j(x) \, dt \, dx \, \frac{d\eta}{(2\pi)^n}$$
$$+ \mathcal{O}(|\lambda|^{-\infty}).$$

We shall then study the asymptotics of each of the integrals to the right. For that we fix a j and suppress the subscript j. We have :

(12.17) $\Phi(t,x,\eta,\lambda) \stackrel{\text{def}}{=} \varphi(t,x,\eta) - x\cdot\eta - t\lambda = tq(x,\eta) + \mathcal{O}(t^2)|\eta| - t\lambda.$

Consider
$$I(x,\lambda) = \frac{1}{(2\pi)^n}\iint e^{i\Phi}\chi(t)a(t,x,\eta)\,dt\,d\eta.$$

If $\lambda \to -\infty$ we have $|\partial_t\Phi| \sim |\eta| + |\lambda|$ and by integrations by parts in t we conclude that $I(x,\lambda) = \mathcal{O}(|\lambda|^{-\infty})$.

Consider next the case when $\lambda \to +\infty$. Let $F(\eta) \in C_0^\infty(\dot{\mathbb{R}}^n)$ be equal to 1 on $\frac{1}{C} \leq |\eta| \leq C$ for some sufficiently large $C > 0$. For $\eta \in \text{supp}\left(1 - F(\frac{(\cdot)}{\lambda})\right)$ we still have $|\partial_t\Phi| \sim |\lambda| + |\eta|$ and hence

$$I(x,\lambda) = \frac{1}{(2\pi)^n}\iint e^{i\Phi}\chi(t) F\left(\frac{\eta}{\lambda}\right) a(t,x,\eta)\,dt\,d\eta + \mathcal{O}(|\lambda|^{-\infty}).$$

We here pass to polar coordinates by putting $\eta = \lambda r\omega$, $|\omega| = 1$, $d\eta = \lambda^n r^{n-1}\,dr\,d\omega$. Then

(12.18) $I(x,\lambda) =$
$$\frac{\lambda^n}{(2\pi)^n}\int_{S^{n-1}}d\omega \iint e^{i\lambda((\varphi(t,x,\omega)-x\cdot\omega)r-t)}\chi(t)a(t,x,\lambda r\omega)F(r\omega)r^{n-1}\,dt\,dr.$$

Here we apply the stationary phase method to the $dt\,dr$ integral. The phase is

$$\tilde{\Phi}(t,x,\omega,r) = (\varphi(t,x,\omega) - x\cdot\omega)r - t = (tq(x,\omega) + \mathcal{O}(t^2))r - t.$$

The unique critical point close to $t = 0$ is given by

$$t = 0, \quad r = \frac{1}{q(x,\omega)}.$$

In fact,
$$\partial_r\tilde{\Phi} = tq(x,\omega) + \mathcal{O}(t^2)$$
$$\partial_t\tilde{\Phi} = rq(x,\omega) - 1 + \mathcal{O}(t).$$

Next we compute the Hessian at the critical point :

$$\tilde{\Phi}''_{t,r}\left(0,x,\omega,\frac{1}{q}\right) = \tilde{\Phi}''_{r,t}\left(0,x,\omega,\frac{1}{q}\right) = q(x,\omega)$$
$$\tilde{\Phi}''_{r,r}\left(0,x,\omega,\frac{1}{q}\right) = 0.$$

Hence $|\det \tilde{\Phi}''| = q(x,\omega)^2$ and the signature of $\tilde{\Phi}''$ is 0. The stationary phase gives then :

$$(12.19)\quad I(x,\lambda) = \frac{\lambda^{n-1}}{(2\pi)^{n-1}} \int_{S^{n-1}} \chi(0) \left(\frac{1}{q(x,\omega)}\right)^{n-1} \frac{1}{q(x,\omega)} d\omega + \mathcal{O}(\lambda^{n-2})$$

$$= \chi(0) \frac{\lambda^{n-1}}{(2\pi)^{n-1}} \int_{S^{n-1}} \frac{1}{q(x,\omega)^n} d\omega + \mathcal{O}(\lambda^{n-2}).$$

Remark 12.4 In the case $n = 1$ we get the improved estimate

$$(12.19)'\quad I(x,\lambda) = \chi(0) \int \frac{1}{q(x,\omega)} d\omega + \mathcal{O}(\lambda^{-2})$$

$$= \chi(0) \frac{2}{q(x,1)} + \mathcal{O}(\lambda^{-n}).$$

In fact, (12.18) becomes

$$(12.18)'\quad I(x,\lambda) =$$
$$\frac{\lambda}{2\pi} \int_{S^0} d\omega \iint e^{i\lambda t(r(q(x,\omega)+\mathcal{O}(t))-1)} \chi(t) F(r\omega) \left(1 + \mathcal{O}(t) + \frac{\mathcal{O}(t)}{\lambda}\right) dt\, dr.$$

Computing the first two terms in the stationary phase expansion we see that the coefficient of the λ^{-1} term is 0.

On the other hand, since $q(x,\xi)$ is homogeneous of degree 1, we have

$$\int_{q(x,\xi)\leq\lambda} d\xi = \int_{S^{n-1}} d\omega \int_0^{\frac{\lambda}{q(x,\omega)}} r^{n-1} dr = \frac{\lambda^n}{n} \int_{S^{n-1}} \frac{1}{q(x,\omega)^n} d\omega,$$

so

$$(12.20)\quad \frac{\partial}{\partial \lambda}\left(\int_{q(x,\xi)\leq\lambda} d\xi\right) = \lambda^{n-1} \int \frac{1}{q(x,\omega)^n} d\omega,$$

and we can rewrite (12.19) :

$$(12.21)\quad I(x,\lambda) = \frac{\chi(0)}{(2\pi)^{n-1}} \frac{\partial}{\partial \lambda}\left(\int_{q(x,\xi)\leq\lambda} d\xi\right) + \mathcal{O}(\lambda^{n-2}).$$

It then suffices to integrate with respect to $\psi_j(x)\, dx$ and add the N integrals in (12.16), to obtain

Theorem 12.5 For $\chi \in C_0^\infty(]-\varepsilon_0, \varepsilon_0[)$ with $\chi(0) = 1$, we have

$$(12.22)\quad \sum_{j=1}^\infty \hat{\chi}(\lambda - \mu_j) = \begin{cases} \mathcal{O}(|\lambda|^{-\infty}), & \lambda \to -\infty \\ \dfrac{1}{(2\pi)^{n-1}} \dfrac{\partial}{\partial \lambda} V_q(\lambda) + \mathcal{O}(\lambda^{n-2}), & \lambda \to +\infty, \end{cases}$$

where $V_q(\lambda) = \iint_{q(x,\xi)\leq\lambda} dx\, d\xi$.

Estimate of the counting function

We introduce the counting function

(12.23) $\qquad N_Q(\lambda) = N(\lambda) = \#(\sigma(Q) \cap \,]-\infty, \lambda])$,

where in general $\#A$ denotes the number of elements of A. In Theorem 12.5, we may choose χ with the additional property $\hat{\chi} \geq 0$. In fact, it suffices to choose $\chi = \psi * \overline{\check{\psi}}$ for a suitable $\psi \in C_0^\infty$. Here $\check{\psi}(t) = \psi(-t)$. We then deduce from (12.22) that

(12.24) $\qquad \#\left(\sigma(Q) \cap [\lambda, \lambda+1]\right) = \mathcal{O}(\lambda^{n-1}), \quad \lambda \to +\infty.$

Writing $N(\lambda) \leq \sum_{j=0}^{[\lambda]} (N(j+1) - N(j))$, with $[\lambda]$ = largest integer $\leq \lambda$, we obtain

(12.25) $\qquad N(\lambda) = \mathcal{O}(\lambda^n), \quad \lambda \to \infty.$

We rewrite (12.22) as a Stieltjes integral :

(12.26) $\displaystyle\int \hat{\chi}(\lambda - \eta)\, dN(\eta) = \begin{cases} \dfrac{1}{(2\pi)^{n-1}} \dfrac{\partial}{\partial \lambda}(V_q(\lambda)) + \mathcal{O}(\lambda^{n-2}), & \lambda \to +\infty \\ \mathcal{O}(\lambda^{-\infty}), & \lambda \to -\infty, \end{cases}$

Put $G(\lambda) = \displaystyle\int_{-\infty}^\lambda \hat{\chi}(\tau)\, d\tau$. Integrating (12.26) and using Remark 12.4 in the case $n = 1$ (giving $\mathcal{O}(\lambda^{-2})$ instead of $\mathcal{O}(\lambda^{-1})$ in (12.26)), we get

(12.27) $\displaystyle\int G(\lambda - \eta)\, dN(\eta) = \begin{cases} \dfrac{1}{(2\pi)^{n-1}} V_q(\lambda) + \mathcal{O}(\lambda^{n-1}), & \lambda \to +\infty \\ \mathcal{O}(\lambda^{-\infty}), & \lambda \to -\infty, \end{cases}$

Here,

(12.28) $\displaystyle\int G(\lambda - \eta)\, dN(\eta) = \sum G(\lambda - \mu_j) = \sum \int_{-\infty}^{\lambda - \mu_j} \hat{\chi}(\tau)\, d\tau$

$\qquad = \sum \displaystyle\int H(\lambda - \mu_j - \tau)\, \hat{\chi}(\tau)\, d\tau = \int \sum H(\lambda - \mu_j - \tau)\, \hat{\chi}(\tau)\, d\tau$

$\qquad = \sum \displaystyle\int N(\lambda - \tau)\, \hat{\chi}(\tau)\, d\tau = 2\pi N(\lambda) + R(\lambda),$

since $\int \hat{\chi}(\tau) d\tau = 2\pi$, where we introduce

(12.29) $$R(\lambda) = \int (N(\lambda - \tau) - N(\lambda)) \hat{\chi}(\tau) d\tau.$$

For $\lambda > 1$, we get from (12.24)

$$|N(\lambda - \tau) - N(\lambda)| \leq C(1 + |\tau|)(\lambda + |\tau|)^{n-1} \leq \tilde{C}(1 + |\tau|)^n \lambda^{n-1}.$$

Since $\int (1 + |\tau|)^{n-1} \chi(\tau) d\tau < \infty$, we have $R(\lambda) = \mathcal{O}(\lambda^{n-1})$. From (12.28), (12.29), we obtain:

Theorem 12.6 *We have*

(12.30) $$N_Q(\lambda) = \frac{1}{(2\pi)^n} V_q(\lambda) + \mathcal{O}(\lambda^{n-1}), \quad \lambda \to +\infty.$$

Finally we return to the operator $P = Q^m$. Put $N_P(\lambda) = \#(\sigma(P) \cap]-\infty, \lambda])$. Since λ is an eigenvalue of P iff $\lambda^{1/m}$ is an eigenvalue of Q, we have

(12.31) $$N_P(\lambda) = N_Q\left(\lambda^{\frac{1}{m}}\right), \quad \lambda \geq 1.$$

Moreover,

$$V_q\left(\lambda^{\frac{1}{m}}\right) = \iint_{q \leq \lambda^{\frac{1}{m}}} dx\, d\xi = \iint_{p \leq \lambda} dx\, d\xi \stackrel{\text{def}}{=} V_p(\lambda).$$

We therefore obtain

Theorem 12.7 *Under the general assumptions formulated at the very beginning of this chapter, we have*

(12.32) $$N_P(\lambda) = \frac{1}{(2\pi)^n} V_p(\lambda) + \mathcal{O}\left(\lambda^{\frac{n-1}{m}}\right).$$

Exercises

Exercise 12.1 (WKB expansions in dimension 1)

Let $V \in C^\infty(I, \mathbb{R})$, where I is an open interval. For $E > \sup_I V$, one looks for approximate solutions of

(1) $$\left((h D)^2 + V(x) - E\right) u = 0 \quad (h > 0 \text{ small})$$

of the form

(2) $$u(x,h) = a(x,h)\, e^{i\varphi(x)/h},$$

where a and φ depend also on E, φ is real C^∞, and

(3) $$a \sim a_0(x) + a_1(x)\, h + \ldots, \quad h \to 0$$

a) Substitute (2) in (1) and group the terms according to powers in h. Show that to "kill" the term in h^0, φ must satisfy the eikonal equation

(4) $$(\varphi'_x)^2 + V(x) - E = 0$$

which gives $\varphi(x) = \pm \int^x (E - V(y))^{\frac{1}{2}}\, dy$.

b) In order to kill the term in h^1, show that a_0 must satisfy the first transport equation:

$$(2\varphi'_x \cdot \partial_x + \varphi''_{xx})\, a_0 = 0$$

and check that the general solution for this equation is

$$a_0 = (\text{const.})\, (E - V(x))^{-\frac{1}{4}}.$$

c) Killing successively the terms in h^2, h^3, etc., show how one can choose $a_1, a_2, \ldots$.

Show that $a_0, a_1, a_2, \ldots$ are uniquely determined if one prescribes $a_0(x_0)$, $a_1(x_0), a_2(x_0), \ldots$ for some $x_0 \in I$.

d) Show that if $a \in C^\infty(I \times [0, h_0[)$ satisfies (3) with a_j constructed in c) (and recall the meaning of (3)) then for every $K \subset I$, $N, k \in \mathbb{N}$

$$|\partial_x^k\big(((h\, D)^2 + V - E)\, u\big)| \leq C_{K,N,k}\, h^N, \quad x \in K.$$

Exercise 12.2 (Quantization condition in the periodic case)

Let $V \in C^\infty(\mathbb{R}, \mathbb{R})$ be a 2π-periodic function and consider $P = (h\, D)^2 + V(x)$ as an unbounded self-adjoint operator: $L^2(S^1) \to L^2(S^1)$ with domain $H^2(S^1)$, where $S^1 = \mathbb{R}/2\pi\mathbb{Z}$. (In other words, we work in the space of 2π-periodic locally square integrable functions.)

a) Show that the spectrum of P is discrete for every $h > 0$.

b) Let φ be as in Exercise 12.1, with $\varphi'_x > 0$ and $I = \mathbb{R}$ (assuming from now on that $E > \sup_{\mathbb{R}} V$). Show that:

$$\varphi(x + 2\pi) = \varphi(x) + \int_0^{2\pi} (E - V(x))^{\frac{1}{2}}\, dx.$$

c) Let $a(x, h) = a(x, E, h)$ be as in Exercise 12.1. Show that at the level of formal symbols $a(x + 2\pi, E, h) = c(E, h) a(x, E, h)$, where $c(E, h) = 1 + c_1(E) h + c_2(E) h^2 + \ldots$ and $c_1, c_2, \ldots$ are C^∞ functions on $]\sup V, +\infty[$.

d) We define u as in Exercise 12.1, and we set $W(u, \bar{u}) = u\bar{u}' - u'\bar{u}$.
Show first that $\partial_x W(u, \bar{u}) = \mathcal{O}(h^N)$ uniformly on every compact set for every $N > 0$.
Show then that $|c| = 1$ at the level of formal symbols.

e) Writing $c(E, h) = e^{ihd(E,h)}$ with $d \sim d_0 + d_1 h + d_2 h^2 + \ldots$ and realizing d as an asymptotic sum, show that for every compact $J \subset]\sup V, +\infty[$ and $h > 0$ small enough, the Bohr-Sommerfeld quantization condition

$$\int_0^{2\pi} (E - V(x))^{\frac{1}{2}} dx + h^2 d(E, h) = 2\pi k h, \quad k \in \mathbb{Z},$$

allows one to define values $E_k(h) \in J$ with the property that for every $N > 0$ there exists $h_N > 0$ such that if $0 < h \leq h_N$ then $]E_k(h) - h^N, E_k(h) + h^N[$ contains at least two eigenvalues (counted with their multiplicity) of P.

Exercise 12.3 (Quantization condition for a potential well)

Let $V \in C^\infty(\mathbb{R}, \mathbb{R})$. We fix $E_0 \in \mathbb{R}$ and we suppose that $\lim_{|x| \to \infty} V(x) > E_0$, and that $\{x \in \mathbb{R}; V(x) \leq E_0\} = [\alpha_0, \beta_0]$, $V(x) < E_0$ for $\alpha_0 < x < \beta_0$, $V'(\alpha_0) < 0$, $V'(\beta_0) > 0$. Let E be in a small real neighborhood of E_0. For x close to α_0, one looks for an asymptotic solution of $\left((hD)^2 + V - E\right) u = 0$ of the form

$$(1) \qquad u(x, h) = \frac{1}{(2\pi h)^{1/2}} \int e^{i(x\xi + \psi(\xi))/h} b(\xi, h) d\xi$$

where ψ, b depend also on E. Let $\alpha(E)$ be the root of $V(x) - E = 0$ close to α_0, and $\beta(E)$ the root close to β_0.

a) Show that there exists a real-valued C^∞ function $\psi = \psi(\xi, E)$, defined in a neighborhood of $\xi = 0$, $E = E_0$, with

$$(2) \qquad V(-\partial_\xi \psi(\xi)) + \xi^2 - E = 0, \quad \psi(0) = 0, \quad \partial_\xi \psi(0) = -\alpha(E).$$

Show also that $\partial_\xi^2 \psi(0) = 0$, $\partial_\xi^3 \psi(0) < 0$.

b) Show that

$$(3) \qquad \lambda(x, \xi) = \frac{\xi^2 + V(x) - E}{x + \partial_\xi \psi(\xi)}$$

defines a C^∞ function near $x = \alpha_0$, $\xi = 0$, $E = E_0$, and that

$$\lambda(-\partial_\xi \psi(\xi), \xi) = V'(-\partial_\xi \psi(\xi)), \quad (\partial_x \lambda)(-\partial_\xi \psi(\xi), \xi) = \frac{1}{2} V''(-\partial_\xi \psi(\xi)).$$

c) We write

(4) $$((hD_x)^2 + V(x) - E)u = \frac{1}{(2\pi h)^{\frac{1}{2}}} \int e^{i(x\xi + \psi(\xi))/h}(\xi^2 + V(x) - E)b(\xi, h)d\xi,$$

and we look for $b(\xi, h) \sim b_0(\xi) + b_1(\xi)h + \ldots$ with $b_0(0) \neq 0$, such that there exists $a(x, \xi, h) \sim a_0(x, \xi) + a_1(x, \xi)h + \ldots$ with

(5) $$e^{i(x\xi + \psi(\xi))/h}(\xi^2 + V(x) - E)b(\xi, h) = hD_\xi(e^{i(x\xi + \psi(\xi))/h}a(x, \xi)) + \mathcal{O}(h^\infty),$$

or more explicitly

(6) $$(\xi^2 + V(x) - E)b(\xi, h) = (x + \partial_\xi \psi(\xi))a(x, \xi, h) + hD_\xi a(x, \xi, h).$$

Show that, in order to solve (6) for ξ near 0, x near α_0, E near E_0, it is sufficient to solve the sequence of equations,

(6.0) $$(\xi^2 + V(x) - E)b_0(\xi) = (x + \partial_\xi \psi(\xi))a_0(x, \xi)$$

(6.1) $$(\xi^2 + V(x) - E)b_1 - D_\xi a_0 = (x + \partial_\xi \psi(\xi))a_1(x, \xi)$$

(6.2) $$(\xi^2 + V(x) - E)b_2 - D_\xi a_1 = (x + \partial_\xi \psi(\xi))a_2(x, \xi)$$

d) For a given b_0, the unique solution of (6.0) is $a_0(x, \xi) = \lambda(x, \xi)b_0(\xi)$. In order to solve (6.1) it is necessary and sufficient that $(D_\xi a_0)(-\partial_\xi \psi, \xi) = 0$. Show that this equation amounts to an equation for b_0, namely $V'(-\partial_\xi \psi(\xi))\partial_\xi b_0 + \frac{1}{2}(\partial_\xi(V'(-\partial_\xi \psi)))b_0 = 0$ and show that the solutions of this latter equation are $b_0(\xi) = \dfrac{C}{|V'(-\partial_\xi \psi(\xi))|^{\frac{1}{2}}}$.

e) For any b_1, we have then a unique solution a_1 of (6.1). Show how to choose b_1 such that (6.2) becomes solvable, and so on.

f) After truncating b outside a small neighborhood of $\xi = 0$, show that u defined by (1) in a neighborhood of α_0 becomes an asymptotic solution of $(P - E)u = 0 + \mathcal{O}(h^\infty)$ and that $u = \mathcal{O}(h^N)$ and similarly for all the derivatives, for $x < \alpha(E) - \delta$, uniformly for all fixed $\delta > 0$ and all $N \in \mathbb{N}$.

g) Show that for $x > \alpha(E) + \delta$

$$u = a_+(x, E, h)e^{i\varphi_+(x)/h} + a_-(x, E, h)e^{i\varphi_-(x)/h}$$

where

$$a_\pm \sim a_\pm^0(x, E) + a_\pm^1(x, E)h + \ldots$$
$$\varphi_\pm(x) = x\xi_\pm(x) + \psi(\xi_\pm(x)),$$

where $\xi_\pm(x) \gtrless 0$ are the critical points of $\xi \to x\xi + \psi(\xi)$.

Show that $(\varphi'_\pm)^2 + V(x) - E = 0$ and that $\lim_{x \to \alpha(E)_+} \varphi_+ = \lim_{x \to \alpha(E)_+} \varphi_-$.

Show also that

$$a^0_+(x,E) = \frac{e^{-i\frac{\pi}{4}} b_0(\xi_+(x,E))}{|\partial^2_\xi \psi(\xi_+(x,E))|^{\frac{1}{2}}}, \quad a^0_-(x,E) = \frac{e^{i\frac{\pi}{4}} b_0(\xi_-(x,E))}{|\partial^2_\xi \psi(\xi_-(x,E))|^{\frac{1}{2}}}.$$

h) Use Wronskians and show that $|a_+| = |a_-|$. (Write first $u = u_+ + c\bar{u}_+$ with $u_+ = a_+ e^{i\varphi_+/h}$, $c \in \mathbb{C}$ and show that $|c| = 1$.)

i) Try to construct a "global solution" of $(P-E)u = 0$ along the curve $\xi^2 + V(x) - E = 0$.

Get the Bohr–Sommerfeld quantization condition

$$2 \int_{\alpha(E)}^{\beta(E)} \sqrt{E - V(x)}\, dx = 2\pi \left(k + \frac{1}{2}\right) h + h^2 r(E,h)$$

where $r \sim r_0(E) + r_1(E) h + \ldots$ is real-valued.

Notice that the left hand side $C(E)$ is the area enclosed by the curve, and that $T(E) = C'(E)$ is the period of H_p on that curve.

Exercise 12.4

We adopt the assumptions and the notation of Chapter 12.

a) Show that if $U(t,x,y)$ denotes the distribution kernel of the operator U, then $U(t,x,y) = \sum e^{it\mu_j} e_j(x) \overline{e_j(y)}$ with convergence in $\mathcal{D}'(\mathbb{R} \times X \times X)$.

b) Deduce that $\int \chi(t) U(t,x,y)\,dt = \sum \hat{\chi}(-\mu_j) e_j(x) \overline{e_j(y)}$ with convergence in $C^\infty(X \times X)$, for every $\chi \in C^\infty_0(\mathbb{R})$, and in particular, that $\iint \chi(t) U(t,x,x)\,dt\,dx = \sum \hat{\chi}(-\mu_j)$.

c) Show that $U(t,x,y) \in I_{\text{cl}}^{-\frac{1}{4}}(\mathbb{R} \times X \times X; C')$, where $C = \{((t,\tau\,;x,\xi)\,;\ y,\eta) \in T^*(\mathbb{R} \times X) \times T^*X\,;\, (x,\xi) = \exp(-tH_q)(y\,\eta),\ \tau - q(x,\xi) = 0\}$. (Also verify that C is a canonical relation.) Verify that for every fixed t, the distribution kernel $U(t,x,y)$ of $U(t)$ belongs to $I^0_{\text{cl}}(X \times X; C'_t)$, where $C_t = \text{graph}(\exp -tH_q)$.

Hint: first treat the case of small t, then use the group property $U(t+s) = U(t)U(s)$ and the result of Chapter 11 on the composition of Fourier integral operators.

d) Using the information about $WF(U)$ that follows from c), show that $U(t,x,x)$ is well-defined in $\mathcal{D}'(\mathbb{R} \times X)$ and that

$$WF(U(t,x,x)) \subset \{(t,\tau\,;x,\xi-\eta)\,;\,(x,\xi) = \exp(-tH_q)(y,\eta),\ \tau = q(x,\xi)\}.$$

e) Deduce that $u(t) = \int U(t,x,x)\,dx$ is a well-defined element of $\mathcal{D}'(\mathbb{R})$ with sing supp $u \subset \{t \in \mathbb{R};\ \exists (x,\xi) \in T^*X\backslash 0,\ (x,\xi) = \exp(-t\,H_q)(x,\xi)\}$. (In other words sing supp u is contained in the set of periods of the H_q-flow.)

f) Show that $\int \chi(t)\,u(t)\,dt = \sum \hat{\chi}(-\mu_j)$, $\chi \in C_0^\infty(\mathbb{R})$ and deduce that $u = \sum e^{it\mu_j}$ with convergence in $\mathcal{S}'(\mathbb{R})$.

Notes

Theorem 12.7 was proved in the case of second-order operators by Avakumovič [Av] and Levitan [Lev1,2,3] who used the associated wave equation on $\mathbb{R} \times X$; in the general case the theorem is due to Hörmander [Hö5] and we have followed his method here (exploiting though the general stationary phase method). The interesting feature of Theorem 12.7 is the remainder estimate which cannot be improved without extra assumptions. The leading term in the expansion, the classical Weyl term, can be and has been obtained by many other methods. Spectral theory is one of the main applications of microlocal analysis; for some further developments see Ivrii [I1], Duistermaat–Guillemin [DG], [Hö4], Chazarain [Ch], Colin de Verdière [Co2], Weinstein [W], Robert [R]. Exercise 12.4 deals with the so-called Poisson formula, observed by Chazarain [Ch], Duistermaat–Guillemin [DG]. Exercises 12.1–3 deal with the 1-dimensional WKB method, which becomes a deep area in the analytic case. (See Grigis [Gr], Voros [V], Candelpergher–Nosmas–Pham [CanNoP].)

Bibliography

[AGé] S. Alinhac, P. Gérard, *Opérateurs pseudo-différentiels et théorème de Nash-Moser*, Interéditions, Savoirs actuels (Paris) (1991).

[Av] V.G. Avakumovič, *Über die eigenfunktionen auf geschlossenen Riemannschen Mannigfaltigkeiten*, Math. Z. 65(1956), 327–344.

[B] R. Beals, *A general calculus of pseudo-differential operators*, Duke Math. J. 42(1975), 1–42.

[BF] R. Beals. C. Fefferman, *Spatially inhomogeneous pseudo-differential operators I*, Comm. Pure Appl. Math. 27(1974), 1–24.

[Bo] L. Boutet de Monvel, *Hypoelliptic operators with double characteristics and related pseudo-differential operators*, Comm. Pure Appl. Math. 27(1974), 585–639.

[CaV1] A.P. Calderón, R. Vaillancourt, *On the boundedness of pseudo-differential operators*, J. Math. Soc. Japan 23(1972), 374–378.

[CaV2] A.P. Calderón, R. Vaillancourt, *A class of bounded pseudo-differential operators*, Proc. Nat. Acad. Sci. USA 69(1972), 1185–1187.

[CanNoP] B. Candelpergher, C. Nosmas, F. Pham, *Approche de la résurgence*, Actualités mathématiques, Hermann, to appear.

[Ch] J. Chazarain, *Formule de Poisson pour des variétés riemanniennes*, Inv. Math. 24(1974), 65–82.

[ChP] J. Chazarain, A. Piriou, *Introduction à la théorie des équations aux dérivées partielles linéaires*, Gauthier Villars (Paris) (1981).

[Co1] Y. Colin de Verdière, *Quasi-modes sur les variétés riemanniennes*, Inv. Math. 43(1977), 15–52.

[Co2] Y. Colin de Verdière, *Sur le spectre des opérateurs elliptiques à bicaractéristiques toutes périodiques*, Comment. Math. Helv. 54(1979), 508–522.

[CF] A. Cordoba, C. Fefferman, *Wave packets and Fourier integral operators*, Comm. P.D.E. (1978), 979–1005.

[Cot] M. Cotlar, *A combinatorial inequality and its application to L^2 spaces*, Rev. Mat. Cuyana, 1(1955), 41–55.

[De] J.M. Delort, *F.B.I.-transformation, second microlocalization and semilinear caustics*, Springer LNM 1522(1992).

[D] J.J. Duistermaat, *Fourier integral operators*, lecture notes, Courant Institute of Math. Sci. (New York) (1973).

[DG] J.J. Duistermaat, V. Guillemin, *The spectrum of positive elliptic operators and periodic bicharacteristics*, Inv. Math. 29(1975), 39–79.

[DHö] J.J. Duistermaat, L. Hörmander, *Fourier integral operators II*, Acta Math. 128(1972), 183–269.

[DS] J.J. Duistermaat, J. Sjöstrand, *A global construction for pseudodifferential operators with non-involutive characteristics*, Inv. Math. 20(3)(1973), 209–225.

[E] Yu.V. Egorov, *On canonical transformations of pseudo-differential operators*, Uspekhi Mat. Nauk 25(1969), 235–236.

[FP1] C. Fefferman, D.H. Phong, *On positivity of pseudo-differential operators*, Proc. Nat. Acad. Sci. USA 75(1978), 4673–4674.

[FP2] C. Fefferman, D.H. Phong, *The uncertainty principle and sharp Gårding inequalities*, Comm. Pure Appl. Math. 34(1981), 285–331.

[Gå] L. Gårding, *Dirichlet's problem for linear elliptic partial differential equations*, Math. Scand. 1(1953), 55–72.

[Ga] H. Gask, *A proof of Schwartz' kernel theorem*, Math. Scand. 8(1960), 327–332.

[Gr] A. Grigis, *Estimations asymptotiques des intervalles d'instabilité pour l'équation de Hill*, Ann. Sci. Éc. Norm. Sup. 4:ème sér. 20(1987), 641–672.

[Ha] N. Hanges, *Parametrices and propagation of singularities for operators with non-involutive characteristics*, Indiana Univ. Math. J. 28(1979), 87–97.

[HS] B. Helffer, J. Sjöstrand, Appendix b in *Semiclassical analysis for Harper's equation III. Cantor structure of the spectrum*, Bull. de la SMF 117(4)(1989), mémoire n° 39.

[Hö1] L. Hörmander, *Linear partial differential operators*, Grundlehren, Springer 116(1964).

[Hö2] L. Hörmander, *Fourier integral operators I*, Acta Math. 127(1971), 79–183.

[Hö3] L. Hörmander, *The Weyl calculus of pseudo-differential operators*, Comm. Pure Appl. Math. 32(1979), 359–443.

[Hö4] L. Hörmander, *The analysis of linear partial differential operators I–IV*, Grundlehren, Springer, 256(1983), 257(1983), 274(1985), 275(1985).

[Hö5] L. Hörmander, *The spectral function of an elliptic operator*, Acta Math. 124(1968), 193–218.

[Hö6] L. Hörmander, *Pseudodifferential operators and non-elliptic boundary problems*, Ann. of Math. 83(1966), 129–209.

[Hö7] L. Hörmander, *On the existence and regularity of solutions of linear pseudodifferential equations*, L'Enseignement mathématique 17(2)(1971), 99–163.

[Ia] D. Iagolnitzer, *Microlocal essential support of a distribution and decomposition theorems – an introduction*, in Hyperfunctions and theoretical physics, Springer LNM 449(1975), 121–132.

[IaSta] D. Iagolnitzer, H.P. Stapp, *Microscopic causality and physical region analyticity in S-matrix theory*, Comm. Math. Phys. 14(1969), 14–55.

[I1] V. Ivrii, *Precise spectral asymptotics for elliptic operators*, Springer LNM 1100(1984).

[I2] V. Ivrii, *Wave fronts of solutions of certain pseudo-differential equations*, Functional Anal. Appl. 10(1976), 141–142.

[I3] V. Ivrii, *Wave fronts of solutions to some microlocally hyperbolic pseudo-differential equations*, Soviet Math. Dokl. 17(1976), 233–236.

[Ke] J.B. Keller, *Corrected Bohr-Sommerfeld quantum conditions for non-separable systems*, Ann. Physics 4(1958), 180–188.

[KSt] A.W. Knapp, E.M. Stein, *Singular integrals and the principal series*, Proc. Nat. Acad. Sci. USA 63(1969), 281–284.

[KN] J.J. Kohn, L. Nirenberg, *On the algebra of pseudo-differential operators*, Comm. Pure Appl. Math. 18(1965), 269–305.

[L] P.D. Lax, *Asymptotic solutions of oscillatory initial value problems*, Duke Math. J. 24(1957), 627–646.

[LN] P.D. Lax, L. Nirenberg, *On stability for difference schemes, a sharp form of Gårding inequality*, Comm. Pure Appl. Math. 19(1966), 473–492.

[Le] J. Leray, *Lagrangian analysis and quantum mechanics, a mathematical structure related to asymptotic expansions and the Maslov index*, M.I.T. Press (Cambridge Mass.) (1981).

[Lev1] B.M. Levitan, *On the asymptotic behaviour of the spectral function of a self-adjoint differential equation of second order*, Izv. Akad. Nauk SSSR Ser. Mat. 16(1952), 325–352.

[Lev2] B.M. Levitan, *On the asymptotic behaviour of the spectral function and an expansion in eigenfunctions of a self-adjoint differential operator of second order II*, Izv. Akad. Nauk SSSR Ser. Mat. 19(1955), 33–58.

[Lev3] B.M. Levitan, *Asymptotic behaviour of the spectral function of an elliptic equation*, Uspekhi Mat. Nauk 26(6)(1971), 151–212, Russ. Math. Surv. 26(6)(1971), 165–232.

[Ma1] V.P. Maslov, *Théorie des perturbations et méthodes asymptotiques*, (translated by J. Lascoux and R. Seneor), Dunod (Paris) (1972).

[Ma2] V.P. Maslov, *Méthodes opérationelles* (translated from Russian), Editions Mir (Moscou) (1987).

[MaFe] V.P. Maslov, M.V. Fedoriuk, *Semi-classical approximation in quantum mechanics*, Math. physics and applied mathematics, Reidel (Dordrecht) (1981).

[MS] A. Melin, J. Sjöstrand, *Fourier integral operators with complex valued phase functions*, Springer LNM, 459, 255–282.

[Me] R. Melrose, *Equivalence of glancing hypersurfaces*, Inv. Math. 37 (1976), 165–191.

[R] D. Robert, *Autour de l'approximation semi-classique*, Progress in mathematics, Birkhauser (1987).

[Sa] M. Sato, *Hyperfunctions and partial differential equations*, Conf. on Funct. Anal. and related topics, Tokyo (1969), 31–40.

[SaKK] M. Sato, T. Kawai, M. Kashiwara, *Hyperfunctions and pseudo-differential equations*, Springer LNM 287(1973), 265–529.

[Sch] L. Schwartz, *Théorie des distributions*, 1, 2, Hermann (Paris) (1957, 1959).

[Se] R.T. Seeley, *Complex powers of elliptic operators*, Proc. Symp. in Pure Math. A.M.S. Providence RI, 10(1967), 288–307.

[Sh] M.A. Shubin, *Pseudodifferential operators and spectral theory*, Springer series in Soviet math., Springer (1987).

[S] J. Sjöstrand, *Singularités analytiques microlocales*, Astérisque, 95(1982).

[St] S. Sternberg, *Lectures on differential geometry*, Prentice Hall (1965).

[T] M. Taylor, *Pseudodifferential operators*, Princeton Univ. Press (Princeton) (1981).

[Tr] F. Treves, *Introduction to pseudodifferential and Fourier integral operators*, vol. 1,2, Plenum Press (New York)(1980).

[V] A. Voros, *The return of the quartic oscillator. The complex WKB method*, Ann. Inst. H. Poincaré, phys. th. 29(3)(1983), 211–338.

[W] A. Weinstein, *Asymptotics of eigenvalue clusters for the Laplacian plus a potential*, Duke Math. J. 44(1977), 883–892.

Index of notations

$\mathbb{R}, \mathbb{R}^n, \dot{\mathbb{R}}^n$	5	S_{cl}^m	35
$\mathbb{N}, \dot{\mathbb{N}}$	5	L_ρ^m	35
$|\alpha|, x^\alpha, \partial_x^\alpha, D_x^\alpha$	5	L_{cl}^m	35
$C^k(X), X$ open	5	S^{n-1}	41
$C^\infty(X)$	5	$u * v$	43
$C^k(X, I), I \subset \mathbb{R}$	5	$(u \mid v)$	44
$C_0^k(X)$	5	$(u \mid v)_{H^s}$	44
$S_{\rho,\delta}^m$	5	$H^s(\mathbb{R}^n)$	44
$S^{-\infty}$	6	$H_{\text{loc}}^s(X)$	44
$x \cdot \xi$	6	$H_{\text{comp}}^s(X)$	44
$g'(x)$	6	$\{a, b\}$	48, 58
$B((x,\theta),\varepsilon)$	9	TX, T^*X, T_xX, T_x^*X	55
$\mathcal{E}'(X)$	10	f^*, f_*	56
$\mathcal{D}(X), \mathcal{D}'(X), \mathcal{D}'^{(k)}(X)$	10	$\wedge$	57
$S_{\rho,\delta}^\infty$	10	d	57
$I(a, \varphi)$	12	$\lrcorner$	58
${}^tA$	14	H_f (Hamilton field)	58
$B(0, 1)$	15	$T_\rho \Lambda^\perp$	60
$\#A$	15	$p^{(j)}, p_{(j)}$	62
$\mathcal{E}'(K)$	15	$\complement$ (complement)	67
$\|u\|_{C^k}$	16	S^k	68
$\mathcal{S}'(\mathbb{R}^n)$	20	$I^{-\infty}$	68
$\mathcal{F}u(\xi) = \hat{u}(\xi) = \int e^{-ix\xi} u(x)\,dx$	20	$S_{\rho,\delta}^m(V)$	77
$\text{sgn}(Q)$	21	$WF(A)$	77
$\mathcal{S}(\mathbb{R}^n)$	27	$WF(u)$	77
$\Delta(X \times X)$	27	$W \subset\subset V$ (for conic sets)	77
$L_{\rho,\delta}^m$	27	Λ_φ	79
$K_A(x, y)$	27	$\mathcal{D}'_\Gamma, \mathcal{E}'_\Gamma$	80
$L^{-\infty}(X)$	28	$WF'(K), WF'_X(K)$	81
σ_A	30	σ (symplectic form)	57, 97
$\mathcal{O}$	30	$\hbar$	99
A^*	32	$I^m(X, \Lambda)$	122
$a \# b$	33	$S_P(x, \xi)$	126
S_ρ^m	35	$N_Q(\lambda)$	138

Index of terminology

asymptotic sum (of symbols) 9
bicharacteristic curve 62
bicharacteristic strip 62, 89
Bohr–Sommerfeld quantization condition 143
Borel's theorem 16, Exercise 1.7
bundle 55
Calderon and Vaillancourt's theorem 50, Exercise 4.7
canonical coordinates 55, 97
canonical 1-form 57
canonical 2-form 57
canonical relation 126
canonical transformation 100
cotangent space 55
cotangent vector 55
Cotlar–Stein lemma 50, Exercise 4.6
counting function 138
counting measure 133
critical set of a phase function 12
Darboux theorem 98, Theorem 9.4
density, $\frac{1}{2}$-density 123
differential form 56
Dirac measure 13
distributions on a manifold 38
Duhamel's principle 73, Exercise 6.2
Egorov theorem 108, Theorem 10.1
eikonal equation 68
elliptic 41
Fourier integral distribution 12
Fourier integral operator (FIO) 13
Frobenius integrability condition 104, Exercise 9.2
Gårding inequality 51, Exercises 4.8, 4.9
generating function 101
Hamilton field 58
Hamilton–Jacobi equations 60
harmonic oscillator 129, Exercise 11.1
hypoelliptic 49, Exercise 4.4
index (of an elliptic operator) 53, Exercise 4.10
involutive manifold 63, Exercise 5.4
isotropic manifold 61
Kuranishi trick 34, 40
Lagrangian manifold 60

Lie derivative 59
linearized vector field 63, Exercise 5.2
Maslov line bundle 125
microlocal parametrix 112
microlocally equivalent (operators) 111
Morse lemma 19
non-characteristic 41
order (of a symbol) 5
oscillatory integral 12
parametrix 42
partition of unity 15
phase function 9
Poincaré lemma 45
principal symbol 35
proper (graph or relation) 28
properly supported (pseudodifferential operator) 29
pseudodifferential operators on a manifold 39
pull-back 56
push-forward 56, 83
singular support (of a distribution) 13
smoothing operator 28
subprincipal symbol 126
symplectic coordinates 97
symplectic space 101
tangent space 55
tangent vector 55
type (of a symbol) 5
vector bundle 55
vector field 56
Weyl quantization 36, 50
WKB expansions 139–143, Exercises 12.1–3